T0222830

Springer Praxis Books

More information about this series at http://www.springer.com/series/4097

"WHAT WILL HE GROW TO?"

Punch Cartoon of 1881. King Coal and King Steam look on the infant Electricity with concern

J.B. Williams

The Electric Century

How the Taming of Lightning
Shaped the Modern World

 Springer

Published in association with
Praxis Publishing
Chichester, UK

J.B. Williams
Offord Darcy
St Neots
Cambridgeshire, UK

SPRINGER PRAXIS BOOKS IN POPULAR SCIENCE

Springer Praxis Books
ISBN 978-3-319-51154-2   ISBN 978-3-319-51155-9   (eBook)
DOI 10.1007/978-3-319-51155-9

Library of Congress Control Number: 2017941080

Cover design: Jim Wilkie

Printed on acid-free paper

This Springer imprint is published by Springer Nature
The registered company is Springer International Publishing AG
The registered company address is: Gewerbestrasse 11, 6330 Cham, Switzerland

Contents

Acknowledgments

It would seem appropriate with the subject of this book that the internet, a child of electricity, should have been such a vital tool during the writing. It was of enormous value in finding and retrieving hundreds of relevant articles from journals published throughout the twentieth century. In addition, many articles are published on the internet alone.

I would also like to thank:

The enthusiasts who run the many online museums to show their collections.

Wikipedia images for many useful pictures.

The librarians of Cambridge University Library for leading me through some of the byways of their collections.

Clive Horwood for his encouragement, Maury Solomon, particularly when it became necessary to split a larger work, and Elizabet Cabrera.

Jim and Rachael Wilkie for turning the manuscript into a book with the cover design and editing.

My referees, Robert Martin-Royle, Dr. Andrew Wheen, Myrene Reyes and Professor Graeme Gooday, for their positive comments which made me feel that the nearly 5 years' work had been worthwhile.

Last, but certainly not least, I want to thank my wife who has, without realizing it, taught me so much about history.

List of Figures

List of Tables

1

Introduction

In January 1882, the prolific American inventor and entrepreneur, Thomas Edison, opened a public electric power station—not in New York or even in America, but at the Holborn Viaduct in London, England. This was some 8 months earlier than the one in Pearl Street in New York which is often claimed as the first. Even the Holborn viaduct station had been preceded by one in the previous year at Godalming in Surrey, also in England. All these were local schemes supplying a small area, but showed progress, as before this an electricity generator had only supplied single buildings.

The power station in Godalming lit only a few arc lamps and some of the new incandescent light bulbs in street lamps, while that at the Holborn viaduct provided lighting for the few shops built along it. The advantage of the site was that the cables could be run along the bridge and it was simple, hence providing a prototype for the Pearl Street station. Even the latter was limited to supplying a small number of customers within just the square mile around it.

However, by the end of the twentieth century nearly every house was full of electrical equipment. These would be made up of switches, plugs and sockets, appliances, kitchen equipment, heating controllers and so on. To do the washing we expect washing machines, dryers, irons. To preserve food, we couldn't do without refrigerators and freezers; to process it mixers, grinders, blenders; and then to cook it, ovens, hobs. Life is unimaginable without electric lights and electrically controlled central heating to keep us warm, or air conditioning to keep us cool.

Outside, there is likely to be a car which, while it may not be "electric", as a minimum uses an electric starter, lights and wipers. As we drive along the roads electric traffic lights control the flow. Or we may well travel on an electric railway to work, or travel up and down in electric elevators, lifts or escalators. In the workplace electric-powered machinery is everywhere.

But in 1900, in all but a tiny minority of houses, there were none of these things; not a single one. It is very difficult for us now to imagine a world without electric equipment everywhere, and yet it has only taken one hundred years. The objective of *The Electric Century* is to examine how we got from there to here, following the links of how a few basic inventions, such as the electric motor, could spawn such a vast range of applications.

© Springer International Publishing AG 2018
J.B. Williams, *The Electric Century*, Springer Praxis Books, DOI 10.1007/978-3-319-51155-9_1

The intention is not to give a "history of technology" but to show how the technology shaped the modern world. These developments had a profound effect on the lives of everyone, and yet the route by which they were achieved is largely ignored. Without electric streetcars or trams, and then railroads and railways, suburbs would not have grown, and our towns and cities would be dense and unmanageable. Without electric elevators or lifts the tall buildings of the modern cityscape would not exist. Without the knowledge brought by high-speed printing that enables mass circulation newspapers and cinema newsreels, universal suffrage would hardly have been be possible.

It was Lenin who appreciated that electrification could be a force for social change. He said: "Communism is Soviet Power plus the electrification of the whole country." While it may not have brought communism, electricity has brought a great leveling out of society, as now everyone has access to the basics of life, heat, light, and mobility. The change had been partly in the direction he expected, but not quite. Maybe part of the trend has been from a collective society to a more individualistic one.

The nineteenth century was still one of horse and muscle power. Yes, steam engines existed, but they were large and inflexible. They powered the railways successfully, and many factories, but there was no simple way to subdivide that power. Hence, the large mill became the norm for employment, and the train for transport. Small factories or even road buses were not very successful. It was the coming of electricity that changed all that, and turned the nineteenth "steam" century, into the twentieth "electric" century.

It is the amount of "power" available that has made such a huge difference to our lives. In 1900, the amount of electricity was very small, and mostly used for shop and street lighting. In Britain, for example, it was the equivalent of less than 8 W per head of population. In 2000, the capacity of the generation stations had risen to 1.2 kW per person, or far more than a hundred times greater. This is for every man, woman and child in the population. On top of this is the power of engines, of gasoline and of gas or oil central heating that has been brought under control on our behalf. Horses are now only for sport or leisure, and muscle power is used so sparingly that many feel the need to use up the excess energy in sport or in the gym.

At the turn of the century most people lived their lives within a very small area of where they lived. If they needed to go anywhere, it was on foot. Horses were for rich people and simply too expensive for the average person, and even the coach services were too dear to consider using for more than that special journey, perhaps once a year. It was the coming of cheap, electrically-powered streetcars, trams and trains that changed all this. Gradually, the systems spread out until everyone could benefit.

So often, as in the case of the telephone, a new invention was taken up by business, and then slowly spread into more affluent houses. The take-up in ordinary homes was remarkably slow. Half the century had gone before more than a quarter of British households had a phone, though large numbers were in use for business. By the end of the twentieth century nearly all homes were connected. Something that was the plaything of the few is now the norm for everyone.

How was all this achieved? Not by "Science" but by engineering and technology. Virtually all the knowledge of the relevant physical laws of nature—the work of Ampere, Ørsted, Faraday, Ohm and so on—was understood at the beginning of the twentieth century. It was its exploitation that took the time and effort. In the end, it is only when the knowledge is turned into real devices that it starts to have an impact and change people's lives.

To follow the story we need to look at the people who made things happen. The twentieth century was characterized by the rise of large organizations so the developments are often hidden behind the corporate façade, but where the individuals can be identified their stories are told. Some of these people will be well-known, but others may well be surprising. There are many unsung heroes who made the vital contribution but received little credit because someone else was better at the publicity.

Marconi naturally features, though he invented little himself. He took the ideas of others and put them together into a system, and made it work. That, coupled with commercial acumen which enabled him to pinpoint the initial markets, led to the development of wireless communication. Even then, he had to share his Nobel Prize. Others, such as Charles Merz, Frank Sprague or Samuel Insull, are hardly known and yet their contribution was great.

We don't want to get bogged down in the sterile arguments about who was the first to invent something. When the time was ripe often a number of people came up with much the same idea. Simultaneous invention is quite common. Four people, who may or may not have known of each other's work, came up with similar designs for an incandescent light bulb within a year or so of each other; although only Edison, and to a lesser extent Swan, are remembered. The interest is in those who developed the technology that ended up actually being used, not in those who went charging up technological blind alleys.

To introduce a major technological change there are three distinct phases: invention, development, and exploitation. It can be argued that Swan and Edison didn't even make the invention; it already existed. What they did do was to develop the idea into a practical device and then (particularly Edison) exploit it in the marketplace. That was their real contribution. Marconi took ideas from other people, but put all these together into a system, identified that ships needed what he had to offer and got a whole industry started.

Often there were enormous lengths of time from an original invention to its real exploitation. The electric drill was a nineteenth-century invention, but it was really after World War II that it came into use for the handyman. The simple addition of a battery to produce a portable device took even longer to come through. Again, it took an immense length of time before refrigerators and freezers were a standard in the kitchen.

The Electric Century looks at how electricity provided light and comfort in our home and, in the form of appliances, eased life. Outside it gave us greater freedom of movement. It was left to its children—electronics, computing and telecommunications—to give us entertainment, communications and computing. That is another story and is treated elsewhere.

Yes, electricity is everywhere, and we can't imagine life without it. To slightly misquote an electricity company advert, it is as though our whole lives run through those wires.

2

Chaotic Beginnings

Electricity is really just organized lightning.

George Carlin

On January 4, 1900 Dr. J. Fletcher Little of Harley Street, London, wrote to *The Times*, as one did in those days when one wanted to complain about something. He had gone further and had already arranged for the President of the Board of Trade to receive a deputation and was canvassing for more people who might want to come with him to complain about the supply failures of the Metropolitan Electric Supply Company.[1]

The Metropolitan Electric Supply Company was one of the myriad small electric companies that had sprung up, putting a generator powered by a thumping steam engine in some back yard and starting to connect those locals who wanted a supply. Its patch was Marylebone, running from the north side of Oxford Street in London and covering the area around Harley Street. It had a number of small generating stations scattered about; one was in Manchester Square to the west and one in Rathbone Place to the east. As an example of the patchwork nature of supply, the south side of Oxford Street was supplied by the Westminster Co., and then came the London Co., and the St. James Co., all of whom were north of Green Park.[2]

The deputation met Mr. Ritchie, the President of the Board of Trade, on 17 January, 1900. The 36 members represented 600 complainants, which was a very considerable proportion of the company's customers. Amongst them were six doctors, a dentist, and an optician, representatives from Middlesex Hospital and numerous shopkeepers. Prominent amongst the latter, Mr. John Lewis who, with his Oxford Street store, was the area's largest ratepayer.

Their charge was that the supply repeatedly failed without warning. This meant that some other form of light had to be hastily obtained, which was not always a simple matter. In some establishments, when the electric lights had been put in the gas fittings had been removed, so they were forced to resort to lamps or candles. This was no laughing matter in the middle of a dinner party, but in a shop full of flammable material it became serious. The hospital had had cases where the power had failed in the operating theater in the middle of important operations.

© Springer International Publishing AG 2018
J.B. Williams, *The Electric Century*, Springer Praxis Books, DOI 10.1007/978-3-319-51155-9_2

Even when there had been a supply the voltage had often been so low that it meant turning on more and more lamps in an attempt to obtain sufficient light. Where three or four lamps had once been sufficient now seven were needed and even those gave an unsatisfactory result. In a local church, the illumination had been so bad that they had had to return to using the gas lights. The complainants had considered taking the company to court but the penalties were not high enough to recompense the commercial undertakings for their losses.

The company had claimed to the Board of Trade that they had solved their problems on December 15, 1899 and there should have been no difficulties after that. The complainants gave details as to why that was not true. What they wanted Mr. Ritchie to do was to introduce a Bill into Parliament so that the electric supply companies would be open to being sued for their failures, like other businesses. Ritchie, a Conservative, had other ideas. He wanted to encourage the setting up of a competitive supplier in the area as he felt that some healthy competition would sort out all the problems. This idea was opposed by the Marylebone Vestry, the strange name for the local council, who didn't want another company "breaking up the streets" to lay cables.

In truth, the Metropolitan Electric Supply Company was still in difficulties and was playing for time. Its customer demand was rising and its backyard generating capacity was insufficient, but it had recognized the situation and had in hand a bold plan to resolve the problem. It wasn't quite as incompetent as many of its customers, particularly Dr. Little, seemed to think But it was laboring under a great many difficulties.

The company had realized that it was impossible to carry on trying to generate in their many small stations in the crowded area they served. They had thus built the first phase of a large power station in Willesden, where there was plenty of space for expansion. They bought an eight-and-a-half acre site bounded on two sides by railways and on the third by the Grand Junction canal (later to become the Grand Union canal). This gave them good access to supplies of cheap coal, essential for powering the station economically.

The first problem was that the generators were unavailable due to an engineers' strike, but they overcame this by obtaining them from the American Westinghouse company. The buildings were erected, the boilers, generators and all the other equipment were installed, and then were ready.

The next task was to transport the electricity to their customers some five miles away. This was where position of the station came into its own, because the supply cables could run along the towpath of the Grand Junction canal most of the way. It was in the last mile or so from there to where the cables needed to go under the streets that they ran into difficulties.

To do this the company had gone to Parliament to obtain the necessary statuary powers. Although opposed by the Paddington and Marylebone vestries, they obtained the powers and immediately started laying the cables. However, these local authorities, having lost the first round, were unhappy about the company laying cables to supply other parishes, and took them to court. The councils were only persuaded to suspend the action by an agreement that the Courts would determine whether the existing powers were adequate and, if not, the company would seek further Parliamentary powers.

Needless to say, the company were also trying to keep their customers happy whilst they fought their way through this morass. Some customers still took them to court and they were fined trivial sums, but this was hardly going to deflect them when they were investing vast amounts in the new system. This was, of course, on top of the actual technical problems of installing equipment and cables and getting this relatively untried technology to actually work.

In the midst of this, the Board of Trade was plowing its own furrow. They still thought that competition was the answer to the problems of supply in the Marylebone area and were encouraging new companies. In April, it passed a Provisional Order authorizing the Marylebone Electric Supply Company to set up in competition to the existing company. This was done despite the opposition of the Marylebone vestry which didn't want another company digging up the streets. In the end it did not pass in Parliament, which was a relief for many, including the existing company.

Dr. Little and his friends were not impressed and agreed with the vestry. He was still trying to persuade the Board that making the company liable for more substantial fines was the way forward, rightly realizing that the current level was not sufficient to deter them. He was also concerned about the high cost of the electricity which, at an average of 10 cents (5 old pence at the time) per unit, he claimed was considerably higher than in other parts of London where it ranged from 9.5 to 8.5 cents.

In the meantime, the company finally got its new Willesden station open with at least part of its customer base connected to the system. It still had a further section around Covent Garden to connect up but that was in hand. It seemed at last that its troubles were nearly over. A next phase was being constructed at Willesden, which should enable it to keep its supply capacity ahead of demand. It was trying to raise more capital so that it had sufficient funds to carry on with its expansion program.

Early the next year, just as things seemed to have settled down, more difficulties arose. The Borough of St. Marylebone (the vestry had become a borough) approached Parliament for powers to get the supply of electricity in its area into its own hands. There was a strong suspicion that the Borough had only become interested in supplying the area with electricity now that the company had at last become profitable.

In reality, it was a totally retrograde step, and driven by pure parochialism, making little economic sense. The Borough had to buy out the assets of the company in the area, and then set up their own local generating station as the Willesden one was not in the borough and so not part of the deal. Before that was completed they had to purchase a "bulk" supply from the company. In the end, it took some years and a great deal of money to finally get the whole thing properly set up.

This is the story of just one undertaking in one area in one city, but it is typical of the trials and tribulations of the infant electric supply industry in many countries. Whilst this example is about a small area in London, and that city may well have been one of the worst, such experiences were repeated elsewhere across the developed world. The *New York Times*, for example, was complaining that though the companies could supply electricity they couldn't do it reliably or cheaply, and they didn't seem to really have it under control.[3]

One of the problems affecting all cities was that the generating stations needed to be local and so were close to where people lived. Inevitably this caused difficulties for the neighbors. There were constant law suits as the residents reached the limits of their tolerance of the smoke, cinders and particularly the vibration from the machinery. In November 1900 in New York alone three cases of this nature came before the courts.[4]

In America, there were a few cities where the population was large enough to really interest the private electricity supply companies. In Boston, New York and Philadelphia, for example, the potential was such that a number of companies were competing for customers in the same areas.[5] Here, rather than digging up the streets, the wires were strung on posts past the users' houses. It meant that the householder could choose their supplier, but at the price that outside in the street there was a jungle of wires.

In some ways, for those who lived in the cities, it was a good time for customers but really it was crazy. Here there was competition, but out in the countryside, where long runs of wire were needed, it wasn't profitable, so people living outside the cities usually didn't have a supply. It was a situation that couldn't last. By the time of the Great Depression, the government had stepped in to rationalize the situation. One supplier in an area became the norm.

Already some of the basic issues surrounding an electricity supply were coming to the fore. Was it just a service which the customer could take or leave, or was it a utility which should aim to supply everyone? It was becoming clear that larger generators were more efficient, but how could the resulting monopoly be controlled? Inevitably, there was going to have to be some regulation of the industry as it grew.

As could be seen from the people who complained about the Metropolitan Electric Supply Company, most of the customers were either shops or other commercial establishments. There were very few private customers, and these tended to be the wealthy. Across the whole of Britain in 1900 there were possibly as few as 50,000 houses connected, representing about two-thirds of one percent of all households.[6] In America, things were a little further advanced and this was perhaps one or two percent.

It is hardly surprising that there were so few. The supply was very patchy, depending on where somebody decided to set up a generating station. In America, the companies just went where they thought they could make a profit. In Britain, as we have seen, it was such a struggle as to discourage most, requiring Parliamentary approval and so on. Then the generation was so inefficient that it was fabulously expensive. The unit cost of the electricity of the Metropolitan Electric Supply Company at 10 cents in 1900 is the equivalent of more than a dollar at the end of the century.[7] We can't imagine paying anything like this amount to light our houses. Charges in 2000 were more like one twenty-fifth of the cost at the beginning of the century.

It was hardly surprising that almost no one had electricity in their homes; far more basic things were missing. Few houses had running water. Most householders had to go to some well or pump or local supply and carry the water home in buckets. The lucky ones might have an outside faucet. Real luxury was an inside faucet in the kitchen. If water was needed anywhere else in the house for the laundry or the very occasional bath, then buckets had to be filled and carried to where it was needed. No wonder most of these activities took place in the kitchen.

With no piped water, running hot water was not even a dream. A tiny number of wealthy people had systems, but these were very complicated. With no electric pumps the systems depended on a "gravity" system where the large diameter pipes were gently angled up from the fire back boiler or furnace so the hot water could rise to where it was needed.

Washday was a nightmare. Quite apart from fetching all the water, the heating was a problem. The lucky ones had access to a "copper". This was a large bowl, usually made of copper, set into a brick structure where a fire could be laid underneath. Given enough time, this would heat the water which had been poured into the copper and could be used to boil the clothes. All steeping, rinsing and so on required the carrying of yet more water. Though some washing machines did exist they could hardly be used without electric power and running water.

Heating in houses was largely by coal. As we saw, the washing water could be heated in the copper, but most was in pans on the "range". This was a cast-iron monstrosity that dominated the kitchen. It was necessary to light the fire early in the morning to have hot water later on, and also so there was heat to cook the breakfast. There was considerable labor in humping coal to keep it running all day. For this reason, though other rooms would often have open grates, they were not often lit because of the labor involved (unless you could afford servants to do it).

All this coal being burnt in the densely-packed cities created another problem—smog. On a day when it was naturally foggy, all the smoke coming out of the thousands of chimneys was trapped under the mist and sat quite close to the ground. It was choking, eye-watering stuff, often so dense that you could barely see your hand in front of your face. In London, these smogs earned the nickname "pea-soupers" because they had a slight green tinge and about the same consistency.

Partly these problems were a function of the density of the city. People lived on top of each other, mainly because with no affordable transport they needed to be within walking distance of their place of work. What transport there was, was horsedrawn and relatively expensive. Railroads were for longer distances and not possible for most people as a means of commuting.

The problem was particularly acute in London. The population density in the center was such that the railroads had been unable to penetrate it, and their terminus stations sat in a ring around the edge. Within that there was no easy way to proceed. There were two approaches. It had been found that laying rails in the street, like for a train, meant that a horse or horses could pull a larger vehicle and the horsedrawn streetcar or tram became popular in the 1870s and 1880s. Though it did help, it was still out of the reach of many people on a regular basis.

The second method was to build underground railroads. In London, the Metropolitan and District companies were both set up to do this. It wasn't truly underground as all they did was to dig a large trench and then bridge it over where streets or buildings needed to be over the top. As much as possible was left open to let the smoke out from the steam engines. The foulness of the atmosphere in the tunnels had to be breathed to be believed. Again, this was relatively expensive method of travel, and not really the solution.

Though the internal combustion engine existed, there were very small numbers of vehicles and motor buses on the roads, so they were not yet able to make any contribution to relieving the congestion. So travel in the city centers really depended on horses, and they produce effluent in large quantities. With horses almost as common as people, the streets smelt like farmyards, and it was a worry that the problem of clearing the streets of the waste would be the limiting factor in the development of cities. Despite apocryphal stories of piles of horse manure many feet high, it never got that bad, but unless something changed the system was sooner or later going to choke itself.

Once people got to work, they mostly used muscle power. Ships were unloaded by cranes and then men humping the produce. Coal, on which the whole economy depended, was dug with picks and shovels and then hauled away by ponies. Fields of grain were cut by scythes, as were lawns, or if machines were available they were hand-powered or drawn by horses.

The only other form of power was steam. In textile mills, large steam engines drove long overhead shafts and lethal belts brought the power down to individual spinning machines or looms. This was practical where large numbers of similar machines were used, but very inconvenient on a smaller scale or where the machines varied widely in size and form. Clearly there was a need for something more flexible.

Electricity supply was still a shambles. London had around nineteen separate organizations, some private and some municipal, while New York had more than thirty companies supplying electricity.[8] The pattern all over was the same—small, local, inefficient operations set up to supply just the local patch, with only some commercial and a few rich customers. There was no real ambition to supply everyone, or to spread it around the countries. It didn't look very promising.

Despite the shaky start there were those who could see the potential. At the end of 1900 the editor of *The Times* in London thundered:

> The application of electricity on the largest scale to the satisfaction of the common wants of life is one of the most important questions with which the public and the legislature have to concern themselves in the present day. Owing in part to a certain slowness in accepting innovations which is characteristic of all of us, and in part to legislation of a confused and discouraging kind, we are behind most other nations in electric lighting, electric traction, and the distribution of electric motive power for machinery.[9]

What was exercising the editor was the obstructionism that was causing delays in getting a new electric streetcar tramway into operation, but even he could see beyond this. Thinking people were beginning to understand that Britain was falling behind in this area, and it can be argued that this was where the pre-eminence of the British Empire began to wane as it fell behind particularly America and to a lesser extent Germany in the exploitation of electricity.

However, everywhere things were stirring. In some quarters there was an expectation of great possibilities in the future, but it didn't seem likely then that electricity would define the century and it would become such an essential part of our lives.

We can now start to follow the specific strands that gradually built up into the network that now encases us.

NOTES

1. This whole saga comes from the pages of *The Times* starting on January 5, 1900 and running right through to December 14, 1904.
2. Poulter J.D. An Early history of Electricity supply. Map on page 103.
3. *New York Times*, December 29, 1900.

4. *New York Times*, November 14 and 17, 1900.
5. Buzz, How electricity grew up? A brief history of the electrical grid, available at: https://power-2switch.com/blog/how-electricity-grew-up-a-brief-history-of-the-electrical-grid/.
6. There are no reliable statistics for this time. This estimate has been made by looking at the number of customers of the various supply organizations and estimating the domestic percentage, and also extrapolating back from the percentage of households connected figure of 1910 in the ratio of the amount of electricity supplied in these 2 years. These two methods give fairly close agreement.
7. Inflation figure from 'Consumer price inflation since 1750', Office for National Statistics.
8. Con Edison, A Brief History of Con Edison, available at: http://www.coned.com/history/.
9. *The Times*, December 29, 1900.

3

Lighting that Doesn't Need Lighting

How many people does it take to invent a light bulb?

In 1802, Humphrey Davy, of miner's lamp fame, built an enormous battery, to Volta's newly-invented form, in the basement of the Royal Institution in London. He took two carbon rods, and connected each one through wires to the ends of the battery. When the two rods were touched together and then pulled slightly apart, a brilliant white light was produced. He had invented the arc lamp. As not everyone could have a large battery in the basement it was another 50 years before it was any practical use.

At the time, light came mostly from candles, but it was gas lighting that took over in Europe and America. From the 1820s, its use steadily increased as more gasworks were built. In 1823, for example, 52 English towns were lit by gas, but by 1859 there were nearly a thousand gasworks.[1] After gas fittings were introduced into the new Houses of Parliament in that year its use spread even more rapidly.[2]

The light output of the jets, burning gas in the open, was 8 or 16 candlepower, which was a considerable improvement on the candles themselves. To get more illumination fittings often had more than one burner, and elaborate arrangements with multiple jets were used in the centers of rooms. These were similar to chandeliers and hence known as gasoliers. To light them was tricky; a match or spill had to be carefully brought to each output once the gas was turned on.

The arc lamp hadn't been forgotten, but what was needed for it to be more widely used was a continuous source of electricity. Michael Faraday's experiments on the relationship between electricity and magnetism in 1831 had led to many attempts to turn mechanical work into electricity. As usual this took a great deal of time and it wasn't until about 40 years later that practical generators started to appear. The way was now open for arc lighting, and a system was successfully trialed at the South Foreland lighthouse in Kent in 1857. Dungeness lighthouse was equipped with arc lamps in 1862.[3]

There was, however, a problem. As the arc burns the carbon rods are eaten away and must be adjusted from time to time to bring them back to the correct distance from each other. This difficulty was largely overcome in 1876 by a Russian telegraph engineer,

© Springer International Publishing AG 2018
J.B. Williams, *The Electric Century*, Springer Praxis Books, DOI 10.1007/978-3-319-51155-9_3

Paul Jablochoff, who had settled in Paris. Up to this time the carbon rods had been placed tip to tip, but he had the simple idea of making them parallel and vertical so that they burnt away equally and so didn't need frequent adjustment.

Within a year, Jablochoff "candles" were being used to light a department store in Paris and the West India dock and part of the Embankment in London. The brilliant light was particularly useful for lighting large open spaces such as markets, stations, and big buildings. In smaller spaces it was just too bright, and there was no easy way to decrease its light output.

The hunt was on for some way of "sub-dividing" the light to produce something more manageable for ordinary street lighting or for use in smaller shops, offices and even homes. This was such an obvious need that many inventors had been attempting to find a solution for some time.

On December 18, 1878 Joseph Swan took a strange object to talk about and to show to his fellow members of the Newcastle Chemical Society in North East England. It consisted of a glass envelope with platinum wires projecting from each end. The glass was sealed with a pip on the side. The contents was a slender carbon rod whose ends were connected to the wires.[4,5]

A month later, on January 17, 1879 he gave another talk in Sunderland and then, on February 3, to 700 people at the Newcastle Literary and Philosophical Society. At these he was able to power the device, and when he did it glowed brightly. At last he had a practical incandescent electric lamp, a goal he had been trying to achieve for nearly 30 years.

It had long been known that if an electric current flowed through a wire, heat was generated. If the wire had a suitable resistance the temperature could reach the point where it started to glow. The higher the temperature the more light was produced, so the race was on to find a way of maintaining it at white heat without burning out.

Swan realized that the secret lay in putting the filament inside a glass envelope and getting all the air out. What contributed to his success was that he used the elegant vacuum pump, recently invented by the German chemist Herman Sprengel. It used drops of mercury in a capillary tube, each of which taking a small amount of the air, eventually producing a high vacuum.

His first lamps were not really practical because the filament he was using was too thick, and so took too a large current at low voltage. What was required was a higher resistance and this needed a thinner filament. It took another 18 months' development, but then at another public lecture in October 1880, Swan was able to demonstrate a technically and commercially viable lamp using a filament made from a carbonized thread of cotton, the method of manufacture of which he promptly patented (Fig. 3.1).

Straight away he lit his own house with his electric lamps. Within a couple of months William Armstrong, who had made his fortune from hydraulic cranes and armaments, installed 45 lamps in his house, Cragside, powered by a water-driven generator. Lord Salisbury and Lord Kelvin and other go-ahead people were also quick to see the opportunities. Swan set up a company to manufacture his design in quantity.

What drove the acceptance of electric lighting was its use in public buildings. The House of Commons was lit by Swan incandescent lamps in 1881, and in the same year the Savoy Theatre had a spectacular installation of a thousand lamps. The next year both the British Museum and Royal Academy installed systems.

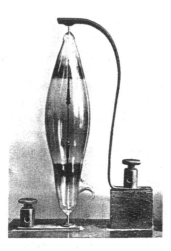

Fig. 3.1 Swan's first incandescent electric lamp. Source: M.E. Swan and K.R. Swan, *Sir Joseph Wilson Swan.*

Swan felt he couldn't patent the basic form of the lamp which had been known for some time, but only the process improvements he had made to the sealing of the glass to the platinum lead in wires and, later, further improvements in this area and the type of carbon used. Finally, in 1883, he came up with a way of making an artificially-extruded thread as the basis of the filament.[6] This became known as "Tamodine".

Thomas Alva Edison, originally from Ohio in the United Sates, was nearly 20 years younger than Swan and by the 1870s already had a formidable reputation as a serial inventor. In 1878, he became interested in the same problem. Being no chemist and someone who worked empirically, he tried a vast number of possible materials for his filament. Eventually, in October 1879, he was in a position to announce a working carbon filament lamp. The following year he had increased the lifetime and produced a satisfactory lamp using a filament derived from bamboo. As with everything else he did, he patented his design.[7]

Swan's increasing success in England soon attracted Edison's attention, who took action against him for infringing his patent. Swan was unworried as, though he had not patented the basic lamp, he could point to the public demonstrations which were clearly before Edison's patent. In patent law this is known as "prior art" and renders a patent invalid if it can be proved. The court agreed with him. Besides, the rivals' designs were quite similar and Swan's patents would make it almost impossible for Edison to manufacture his lamps. Edison soon realized the position and agreed to a deal to form a joint organization, The Edison and Swan United Electric Light Company Limited, to produce the lamps (later using the trade name Ediswan). Swan, faced with this brash American and the possibility of endless litigation, thought this was the better option.

Over the next few years they were in the strange position in Britain of trying to defend Edison's patent while pretending that Swan didn't really have "prior art". Somehow they managed to get away with it, but the price Swan had to pay was that Edison could claim that he invented the light bulb.

Several other people, including Hiram Stevens Maxim, of machine gun fame, and St. George Lane Fox, had come up with designs but Edison's commercial muscle overpowered all of them, so maybe Swan made the right decision. Soon their joint company was manufacturing carbon filament lamps in considerable numbers. However, in America where Swan's patents didn't apply, Edison had a free run and soon the Edison Electric Light Company had conquered the majority of the market.

In some ways this is surprising as the lamps weren't really very good. By modern standards they were very dim. Though the early lamps came in many different outputs they soon settled down to the 8 or 16 candlepower lamps. To get the output of just eight candles required about 33 W of power. To put this in perspective, the incandescent bulbs at the end of the century produced about one candlepower per watt, or four times as much.

The problem was quite simple. If a filament was run at a higher temperature, then more light was produced but it simply burnt out. As with all filaments, there is a trade-off between light output and lifetime. The carbon filament couldn't be run at a high-enough temperature to get a really useful amount of light. The obvious question is why both Swan and Edison settled on carbon filaments. The answer is that they could make them, and at a reasonable price.

Considering that the lamps only produced half the light of a gas jet, and were much more expensive to run, why would anyone want to use them? It was necessary to install wiring and somewhere needed to be found to put the generator, and the thumping steam engine with its boiler, to drive it. This compared with a gas supply that was usually already piped in, or the simple use of portable lamps or candles.

The answer lay in dirt and danger. When the hydrocarbons in the gas burnt (the same was true of candles or lamps) the hydrogen combined with oxygen to form water and the carbon, if completely burnt, formed carbon dioxide. In practice, the burning was never perfect and considerable amounts of carbon were produced and deposited around the room. Also the gas was often not very pure and other noxious substances were often produced. The result was that furnishings, walls and ceilings had to be cleaned or renewed often. For a museum or art gallery this was even more serious as the exhibits could be damaged, and they were some of the earliest users of electric light.

The obvious danger is from the use of a naked flame. One of the things that drove the uptake of gas lighting in public buildings was that fire insurance premiums were reduced. This presumably was due to the fact that, unlike candles or lamps, the burners couldn't be knocked over. Electric lamps without the naked flame were even safer in this respect, and as numerous theaters had burnt down, this undoubtedly was a factor in their adoption.

There was an even more subtle danger with gas lighting. All the time it was lit, carbon dioxide was being pumped into the room. Because the burners tended to be larger, and users wanted more light, this was a greater problem than with oil lamps or candles. Once the carbon dioxide concentration gets up to around one percent it is sufficient to give people headaches and make them feel very ill. In Victorian times this was recognized, and known as the stupefying effects of the gas. I wonder how many of those ladies who passed out did so because of the gas and not just on account of their tight stays? If the gas was not properly burnt, the even more dangerous carbon monoxide could be produced, with fatal results.

For street lighting, none of this was really a problem, and gas was used extensively. Some areas did use electric lamps, despite the increased costs, because they could all be turned on at the flick of a switch rather than the lamplighter having to light each one individually.

Carl Auer Freiherr von Welsbach was an Austrian scientist who had the knack of turning his inventions into commercially-successful products. He is best known for his work on rare earth elements, which led to the development of the gas mantle. A mantle is an impregnated cotton gauze bag which burns away when first lit, leaving a structure made up of the impregnation which will glow brightly in a gas flame.

He first produced a device in the 1880s, but it wasn't very successful, but by 1891 he had perfected and patented an improved device which used 99% thorium oxide with 1% cerium oxide.[8] Von Welsbach rapidly put it into production, and its use spread across Europe. Here was something with which the gas light could fight back, as it produced something like three times the light for the same amount of gas, and was a brighter white. Its drawback was that it was very fragile, and the slightest error with the match or spill when lighting it would result in damage or destruction. The other drawbacks of gas lighting were still there.

Nevertheless, gas dominated the lighting market in 1900. Gas was estimated to have 82% of the market, oil lamps 15%, candles 1% and electricity around 2%,[9] which showed that it was used more in public buildings and street lighting than in homes. The oil lamps, using kerosene or paraffin, were mostly being used in places where the gas mains hadn't reached. It looked as though gas had won the war, but many still believed that the future lay in electricity.

Three things were holding back its development. Unlike gas, which had a lead of 50 years or more, it lacked the infrastructure to supply the customers. It had rapidly been understood that private basement generators weren't the way forward, but it was the mid- to late-1880s before supply organizations, such as Edison or the Metropolitan Electric Supply Co., began to appear and connect up anyone in their area who wanted a supply. Even so they were thinking small. They were only licenced for these tiny areas, and the technical approach taken by most wouldn't allow them to expand more than about a mile from the generator. Money could often be made by just supplying this relatively cushy customer base of wealthy users and not trying to expand further which required more money and effort.

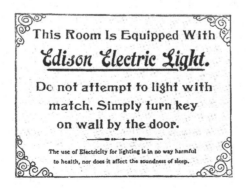

Fig. 3.2 People using early electric lighting had to be taught to switch it on and not try to light it with a match. Source: A. Byers: *The Willing Servants*.

The next problem was the price charged for the electricity. One factor was the inefficiency of the generation. It took in the order of 5–10 pounds weight of coal to generate a unit of electricity. By World War II, this was down to around one-and-a-half pounds.[10]

In addition, there was a more subtle factor. Virtually all the electricity was used for lighting. Lights are little used throughout the day or in the middle of the night, but there is a very sharp peak in demand in the evening. Unlike its competitors, electricity cannot be stored. Coal can be piled into heaps, oil poured into tanks, and gas collected in the large gasholders that used to dominate so many towns. As a result, the system had to be large enough to cope with this peak, but most of the time it was almost idle. The amount actually generated, compared with the theoretical amount if the plant ran at full capacity all the time, was known as the "load factor". For most organizations this was very low, of the order of ten percent,[11] which meant that the plant was being used very inefficiently.

The other problem was that once the gas mantle came into common use around the turn of the century, the electric lamps needed to be improved, both in brightness and in the amount of electricity they used for the light they produced. The way to tackle both these issues was the same; the temperature at which the filament ran needed to be raised, and this required a better material.

Von Welsbach patented a lamp using an Osmium filament (melting point 4900 °F, 2700 °C) in 1898, which came into use in the new century. Though it was an improvement it was difficult to make and not the complete answer. Siemens & Halske in Germany tried tantalum (melting point 5400 °F, 2996 °C), but again it was a difficult and expensive material to use.

In 1906, the American General Electric, GE, which derived from the company set up by Edison, patented a lamp with a tungsten (melting point 6200 °F, 3410 °C) filament. Tungsten is a difficult material to work with and it wasn't until around 1911 that the manufacturing process was finally mastered. The lamp was so superior that its use spread rapidly, despite it initially being sold at a considerably higher price than the carbon filament lamps.[12]

Considering that in the early years of the century only a few percent of homes were wired for electricity and hence had electric light, it might seem surprising that 50 years later the position would be reversed and only a few percent didn't have electric lighting. However, by 1920 the falling cost of electricity and improvements in the lamps meant that the cost of electric lighting fell below that of gas.[13] Now that it had a clear advantage, houses were wired up just so that their occupants could have it.

Gas lighting hung on and was still being used up to World War II, but it had largely disappeared by 1950. It remains even today in odd niches where its nostalgic glow is of interest to tourists; in London, outside the National Gallery in Trafalgar Square, for example, there is one with gas jets, without even a mantle.

There have been a number of improvements to electric lamps since. Irving Langmuir at GE in the USA pinned down the cause of blackening of the glass, and he introduced the inert gas filling which overcame the problem.[14] He also found that coiling the filament increased its performance by around 20%. Beyond those changes and the introduction of frosting, the lamp largely remained unchanged for most of the century, and provided the mainstay for all that time. There has been a tendency to higher light levels, the original 25 and 40 W bulbs rapidly being overtaken by the 60, 75 and 100 W lamps, but the essence is the same. One of the pillars of modern life was in place.

However, the incandescent lamp wasn't to have it all its own way. There is another class of light sources: gas discharge tubes. These come in a number of forms and have steadily gained ground as technical problems have been overcome. These all stem from the work of people such as Herman Geissler in the middle of the nineteenth century when it was discovered that an electric current could be passed through a tube of small quantities of some types of gases and they would glow with a distinctive color.[15]

These were just curiosities used for entertainment until in 1901 an American, Peter Cooper Hewitt, patented the low-pressure mercury vapor lamp.[16] This produced a very bright light, but unfortunately not only was it an unpleasant bluish-green but it also produced large amounts of ultra-violet which could affect people's eyes. The lamp found some niche markets but was not very successful.

In 1910, Frenchman Georges Claude demonstrated the neon tube, with its distinctive red color. By using other gases, or a mixture of them, a range of colors could be produced. The tubes could be bent into a variety of shapes and became very popular as shop and advertising signs, particularly in America during the 1920s and 1930s.[17]

In the early 1930s, experiments with other fillings led to the low-pressure sodium lamp. At first sight this didn't appear very useful as it produced a very narrow spectrum of color in the yellow region. However, as all the energy was concentrated there and this color is close to the point where the eye is most sensitive, the effective efficiency is very high.[18] When sodium lamps were commercially introduced in 1932 there began to be an interest in them for the lighting of major roads.

Apart from the efficiency, which meant a lower running cost for the same output, they had some other advantages. As photographers know, black and white photographs have a sharper contrast than color ones. Though these lamps produced yellow light and not white, it has the same advantage in that the eye only has to distinguish between the single color and black. This means that shapes and movement can be more easily distinguished in the low lighting levels at night. In addition, the single color removes dispersion in mist or fog, so making it easier to see under those conditions.

These advantages led to the yellow sodium lamps steadily spreading along the major roads, giving that characteristic color associated with driving at night. Only much later, in the 1960s and 1970s, were high-pressure sodium lamps produced. These gave an output with a wider spectrum enabling colors to be recognized. The price for this was a reduced efficiency. However, they have become popular and find uses varying from helping plants grow to lighting public spaces such as football stadia, as well as the roads.

Meanwhile, a considerable number of researchers had not given up on the mercury vapor lamp. The basic idea was to coat the inside of the tube with a phosphor which would convert the ultra-violet energy into visible light. The phosphor would fluoresce and so the fluorescent lamp was born. A number of patents[19] were taken out for aspects of this but it was not until 1938 that GE of America and British Thomson Houston in Britain[20] were offering fluorescent tubes for sale.

Gradually, from the 1940s onward they began to find use, particularly in offices and stores where their higher efficiency, some four times that of an incandescent bulb, and hence lower running cost, were important. For these purposes they took over almost completely. They became the ubiquitous "tube" lights, but in commercial buildings they could easily be hidden behind translucent panels. They were never very popular in the home: s they took a time to start; they produced a whiter light that was felt to be less cozy than the yellowish tinge of the incandescent bulbs; and they were difficult to shade in any way.

The question remained: could a lamp be produced that was as simple to use as an incandescent bulb and fitted into the normal socket, but was as efficient as the fluorescent tube? There were considerable difficulties in the way of this. For a reasonable light output the tubes were much too big and would need to miniaturized. Like all gas discharge tubes they needed arrangements to start them, and some form of "ballast" to control the current that flowed. These components were quite large.

This didn't stop people trying, but it wasn't until the 1970s that solutions were found and a series of patents were taken out for various designs.[21] However, they all foundered on the high cost of manufacture or the investment necessary to do this. Then in 1980 Philips launched the model SL, but still with a magnetic ballast like the conventional fluorescent tubes. It wasn't the complete answer.

Another 5 years elapsed before Osram started selling its model EL lamp, which used an electronic ballast and starting system for the first time. This was the way forward as it could be made small enough to fit into the base of the lamp. By suitable coiling or folding of the tube, together with high-performance phosphors, a practical lamp could be made. Within the next few years, a number of other manufacturers also produced similar products. In addition to producing something like five times the light for the same energy consumption, they had another advantage in that their lifetime was much longer.

Initially their cost was rather high due to the investment in new manufacturing plant necessary to make them, but gradually the prices started to fall. Then in the 1990s the world started to become worried about the amount of carbon dioxide being pumped into the atmosphere. Electricity suppliers and governments grabbed at these lamps as a way to reduce the twenty percent of the energy consumed in the home which was used for lighting.[22] As they became more widespread, the rate of increase of energy consumption fell and then the actual energy used is projected to decrease into the twenty-first century as incandescent lamps are phased out.

So, surprisingly, after a century of dominance, the incandescent lamp had to give way to a sophisticated offspring of the gas discharge tube. This is a result that would have been thought ridiculous at the beginning of the twentieth century, and difficult to predict until quite recently. However, it isn't the end of the matter as other even more advanced methods of producing light, such as Light Emitting Diodes (LEDs), are waiting in the wings.

One thing we can be sure of is that the desire for high levels of lighting at any time of night or day, at the flick of a switch, won't go away. The average amount of light that each person uses has increased by almost 80-fold during the century.[23] At the beginning it was provided by gas or lamps, but they have been swept away by electric lighting. It includes street and commercial lighting as well as that in the home, but it is still a staggering increase.

What this means in reality is that, though higher brightness is the norm, now everyone has the same lighting. It is no longer a plaything of the wealthy. It has become part of normal life for everyone.

NOTES

1. Derry T.K. and Williams T.I. *A Short History of Technology*, p. 513.
2. Taylor J. Lighting in the Victorian home, available at: http://www.buildingconservation.com/articles/lighting/lighting.htm.
3. Derry and Williams, p. 630.
4. Swan M.E. and Swan K.R. *Sir Joseph Wilson Swan*, p. 63.
5. Green M. *The Nearly Men*, p. 133.
6. See Swan's British Patents 4933 (1880), 4202 (1881), 5978 (1883). Note that British patents started at 1 again for each year at that time.
7. Kennelly A.E. Biographical memoir of Thomas Alvar Edison, National Academy of the USA, Biographical Memoirs, Volume XV Tenth Memoir.
8. Auer von Welsbach-Museum, Biography, available at: http://www.althofen.at/AvW-Museum/Englisch/biographie_e.htm.
9. Fouquet R. and Pearson P.J.G. Seven centuries of energy services: The price and use of light in the United Kingdom (1300–2000), *The Energy Journal, 27*: 1, 2006.
10. Hannah L. *Electricity* before Nationalisation, p. 432.
11. Hannah, p. 20.
12. Andrews F. *A short history of electric light*, available at: http://www.debook.com/Bulbs/LB07inc1900.htm. Frank Andrews is a knowledgeable collector of early electric light bulbs.
13. Fouquet and Pearson.
14. Bowen H.G. *The Edison Effect*, p. 33.
15. Wikipedia, Geissler Tube, available at: http://en.wikipedia.org/wiki/Geissler_tube.
16. Bellis M. The history of fluorescent light, available at: http://inventors.about.com/library/inventors/bl_fluorescent.htm.
17. Wikipedia, Neon lighting, available at: http://en.wikipedia.org/wiki/Neon_lighting.
18. Hooker J.D. The low pressure sodium lamp, available at: http://www.lamptech.co.uk/Documents/SO1%20Introduction.htm.
19. E.g. US patents 2,182,732 and 2,259,040.
20. Bellis M. The history of fluorescent lights, available at: http://inventors.about.com/library/inventors/bl_fluorescent.htm; Tweedie A.I., *BTH*, available at: http://www.gracesguide.co.uk/BTH.
21. Smithsonian, Compact Fluorescent, The challenge of manufacturing, available at: http://americanhistory.si.edu/lighting/20thcent/invent20.htm#in4.
22. Energy Saving Trust, *The Elephant in the Living Room*.
23. Fouquet and Pearson.

4

Streetcars, Subways, Trains and Suburbs

The essence of the suburb was that it would provide the benefits of country living with the convenience of being close to the city, so that a man (and it was mainly men) could travel by train to work in the noise and grime of the city but come home in the evening to his family and the healthy air of the countryside.

David Dimbleby, *How We Built Britain*

Late at night on September 7, 1888, in Richmond Virginia, 22 electric cars were drawn up nose to tail on a short section of street railway track.[1] At a signal the first one moved off and once it was clear, the next followed it and so on until the whole lot were in motion. It seemed a curious charade, and there was probably some cheating going on as the earlier ones shut off power before the last ones started to draw it.

Nevertheless, the scene impressed the two witnesses from the West End Street Railway of Boston, Massachusetts, which was at the time the largest transit operator in the United States. It was the final step in convincing them that the problems of applying electric traction to trams or streetcars had been solved. It was going to have an enormous impact.

The man running this demonstration was Frank Julian Sprague, an entrepreneurial engineer who, after service in the Navy, had worked for Thomas Edison for a while. They parted company because Edison wanted to concentrate on electric lighting while Sprague was more interested in the possibilities of electric power. He left to form his own company and made significant advances in the design of electric motors. This led him to examine the possibilities of applying them to streetcars.

The system in Richmond had been built in an astonishingly short time after he gained the contract in 1887. By February of the following year the first streetcars were running.[2] In the first week they moved more than 18,000 passengers, but by June, when the whole system was operational, they carried up to 70,000 people per week. Though there had been some difficulties these were steadily overcome. There were stories of broken-down cars having to be rescued by mules, but Sprague denied this.

Sprague had to solve a whole series of problems, but he started with the advantage of his superior motor. The next was how to provide power to the vehicle. He opted for overhead wires as they were the most economic solution and used a "troller" to run along the

© Springer International Publishing AG 2018
J.B. Williams, *The Electric Century*, Springer Praxis Books, DOI 10.1007/978-3-319-51155-9_4

wire to make a connection with it, and the return was through the running rails. From this came the name trolley car or trolley bus for these sorts of vehicles. As part of the contract he had to provide the power station. Sensibly he bought boilers, steam engines and generators from other companies and put them together.

The two visitors from Boston were convinced and ordered a similar electric system from Sprague. This was the beginning of an enormous rush to electrify the street railroad systems, mostly in the USA. Within 2 years Sprague had contracts to construct 113 electric street rail systems, and by 1895 almost 900 electric street railways and nearly 11,000 miles (18,000 km) of track had been built. Within a decade, horsedrawn streetcars had virtually disappeared from America's cities (Fig. 4.1).[3]

In the nineteenth century, as the population became more urbanized, horse transport was clearly in need of improvement. The first step was from around 1870 onwards when the idea of the streetcar took hold. It was a simple matter to combine the advantages of the railroad, running on tracks, with that of a vehicle that ran along the street. This combination produced the horsedrawn streetcar. The downside was that tracks had to be laid in the streets, but the advantages were in a smoother ride and that the horses could pull a larger number of passengers.

However, the problems revolved around the horses. The Boston company, for example, had 8000 horses in 1888. They could only work for a few hours a day and so it took five to seven horses to maintain a service on just one streetcar. With a working life of only 4–5 years, and the expense of feed and stabling, the horse costs ate up 50% of the revenue.[4]

Fig. 4.1 Opening the electric tram system in Bristol, England. Source: http://www.swehs. co.uk/tactive/sparkhome.php

In some places anything up to four horses at a time were needed to pull the streetcar. Thus the fares had to be quite high (Fig. 4.2).

The search was on for a better form of propulsion. The obvious one to try was steam. Locomotives were disguised as streetcars and attached to the front of the normal car. It worked, but the passengers, particularly on an open deck, were likely to get on clean and off again very dirty. In addition, the snorting monster frightened the horses of the other road users; always a serious offence in Victorian times. They never became popular, and a number of companies operating them went bankrupt.

There was a brief vogue for cable hauling. Here, a fixed station powered the system, winding a cable that ran in an underground duct along the length of the tramway. The streetcar was attached to the cable and was hence hauled along. The system could work quite well in ideal circumstances, particularly where there were hills, such as in San Francisco, but it was expensive to build and difficult to maintain.

Ever since it had been realized that not only did a wire moving in a magnetic field generate electricity, but the opposite was also true, people had experimented with attaching electric motors to vehicles. It seemed to hold promise. The first obvious thing to try was to have a set of batteries in the streetcar to power the motor. These could be charged at the terminus and the streetcar could then run up and down to provide the service. With the batteries available at the time the distance was strictly limited, and not satisfactory to provide a service. The weight was excessive and in addition the fumes from the sulfuric acid in the batteries tended to alarm the passengers.

Frank Sprague didn't invent the electric streetcar, but he developed it into a practical system by solving a whole series of outstanding problems. There had been a series of attempts before this, many of them demonstrators at exhibitions, but no one had managed to produce a satisfactory commercial system. Sprague opted for the overhead wire method

Fig. 4.2 A double deck horse streetcar or tram; this one operating in London. The top deck is open and the driver has no protection from the elements. Source: https://en.wikipedia.org/wiki/Trams_in_London#/media/File:London_Tramways_Horse_tram.jpg

to power his cars. He argued that, despite the disadvantage of the unsightliness of the overhead wires, which needed to be strung along the whole length of the route, it was the best system.

Other people had tried running the power supply in or under the ground. Power could be run either in the running rails or on a separate rail alongside. This was unsuitable for street running as there was a danger of losing passengers or any passing people or horses to electrocution. It was thus only suitable where a separate track could be used. This later became important when railways were electrified.

Another method was the underground conduit. A "plow" underneath the tramcar ran in a slot in the ground which contained the electric conductor. It was both safe and neat, but suffered from a number of difficulties. The seaside resort of Blackpool in England adopted it for its early streetcar system along the promenade, but on wet days the conduit filled with sea water, and on dry ones with sand. After a while, the wet sand would short out the electricity supply. It was a nightmare to keep clear and was soon abandoned for an overhead system. In London, the London County Council adopted this system because of opposition to overhead wires, but the heavy costs of maintaining it were a constant financial burden.

One of the attractions of the electric system was the very different way in which the costs stacked up. With horse trams the high cost of care of the horses meant that there was very little that could be done to reduce the fares, which in Britain ranged from 4 cents (two old pence at the time) down to sometimes as low as 2 cents per mile,[5] and hence they were too expensive for most people. The cost of running the power station to provide the power for the electric streetcars was much lower. For example, in Bristol in England, the running costs of the St. George plant powering the Kingswood line was only around 15% of the revenue.[6] There were high installation costs for the overhead wiring, the power station and the streetcars, but the management could juggle with repayment periods so that there was scope to reduce the fares.

In addition, the electric streetcar was faster. Horses only achieved a maximum of 4 miles per hour while the electric streetcar could do much more, sometimes 12 miles per hour (although often local regulations limited their speed to 8 miles per hour). This was a piece of obstructionism which negated some of the potential advantage. Even so, at twice the speed and able to carry more than twice the number of people the streetcar could move far more fare-paying passengers in the day. The reduction in costs was reflected in lower fares. The greatly increased use wasn't just because it was cheaper; it was more attractive and this brought in the passengers. It changed from being the preserve of the better off to something everyone could afford to use.

In Britain, the 1870 Tramway Act had placed very severe restrictions on operators and had given the local councils the right to buy out any private systems after 21 years. This had acted as a break on the development of electric streetcars as the private companies had been reluctant to invest knowing that their licences would soon end. Despite this, some towns and cities introduced them. Sometimes these were operated by companies under licence from the councils, but often by the municipality themselves.

The net result was that the introduction of electric streetcars in Britain was behind that of other industrialized counties. In 1900, Britain had 572 miles of electric tramway,[7] Germany 1800 miles,[8] while in America virtually all had been converted and there were

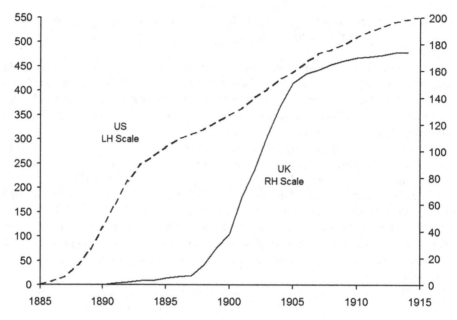

Fig. 4.3 The dash to open electric tramway systems. The US has an earlier start but then slows down while the UK follows the classic 'S' curve. Source: Author.[9]

15,000 miles (Fig. 4.3).[10] By 1902, over half of Europe's total tramway mileage was in Germany. However, things were about to change in Britain and following the more flexible 1896 Light Railways Act, a mad scramble ensued to introduce systems with more than a hundred appearing in the next 10 years.

Suddenly, the average person had a satisfactory means of transport which could be used for commuting as well as weekend trips to parks, countryside and football matches. In many places fares fell further. On Liverpool trams the cost dropped to less than 1 cent instead of the 3 cents per mile on the horsedrawn ones.[11] In some areas the workman's fare was as low as ½ cent per mile. Naturally, the number of passengers leapt and was often three times what it had been.

In Britain, most of the early development took place in smaller towns and cities, and it was only well after 1900 that London got its first electric streetcars. However, they soon became all the rage there. In 1903, electric ones became fashionable as well as popular when the Prince of Wales attended the opening of London County Council's electric tramway and was said to have bought the first ticket with specially minted halfpenny (1 cent).

Most countries, including Britain, had a fare structure that depended on distance, and in practice this was in "stages" at a fixed cost. Steadily these were lengthened for the same money or the prices decreased. They dropped well below 1 cent per mile reaching as low as 5½ miles for 2 cents in some cases by 1907.[12]

In America, however, there was a different arrangement where typically a whole ride cost 5 cents.[13] This was fine for the operators when the tracks only ran for a mile or two in the city centers, but as the networks were extended out into the suburbs it meant that the revenue per mile was reduced. At first, when there were many operators, it meant that the traveler had to pay again when they changed streetcar. After many of these operators were brought under single ownership, the fiction of separate companies was maintained to continue this practice.

Many of these new owners were very dubious characters, like Charles Tyson Yerkes (rhymes with turkeys). He built up his streetcar empire in Chicago by openly bribing the city council to minimize the regulation so that he could make considerable profits.[14] Needless to say, he was hated in the city and eventually this led to his downfall.

The whole thing meant that the typical fares in America were higher than elsewhere, particularly Britain. This is surprising as in Britain they were mainly in private ownership, whereas elsewhere they were often run by the local metropolitan authority. With the greater affluence the companies could get away with this, but the relatively high cost of public transport and the mass production of automobiles contributed to their faster adoption in America.

Even before the fares had significantly reduced, it was clear that the introduction of electric power was a game changer. The streetcars were clean, quiet (relatively) and faster which meant that a more regular service could be provided. It also meant that journeys from further out of town were practical for everyday activities, such as going to work.

This led to the building of "streetcar suburbs", or middle-class areas. In Philadelphia, the various lines opened up the development of much of Delaware County, eastern Montgomery and southern Bucks.[15] In Camden County, New Jersey, it was the Public Service Railway that enabled the development of suburbs such as Collingswood and Haddonfield.

In Boston, the extensions of the streetcar system allowed the development of the suburbs such as Wellington Hill, where people could live and make the six-mile commute which would have been impractical before.[16] Neighborhoods such as Brighton and Somerville had the same treatment. Wherever the streetcars extended out from the centers of cities there was the same phenomena. It soon reached the point where the railway companies became involved in these building developments. In some cases, that was more profitable than running the streetcars.

In Britain, with the later start, even before the tramways opened the opportunity had been spotted. "To Capitalists, Builders and Speculators" ran the advert for building land near Kingswood, Bristol, emphasizing that it was close to the tramway that was soon to open. The way was open for a greater proportion of the population to consider moving out of the center of the city. By World War I, building had taken place all along the streetcar line, effectively creating continuous suburbs along these routes.[17] The horsecars had had no such effect. In towns and cities across the country the same occurred. More and more people were taking the opportunity of efficient transport to move out of the centers.

On the outskirts of London, considerable development took place along the new streetcar lines. Hendon, Finchley and the strip from Wood Green to Enfield all grew enormously between 1900 and 1914 as the electric tramways opened up the area.[18] Speculative builders were always on the look-out for attractive locations on a streetcar route. Tramways were also considered an important factor in the siting of large-scale local authority housing projects such as the London County Council development in Tottenham at the White Hart Lane estate.[19]

In the late nineteenth century, London was by far the biggest city in the world. It was thus far more congested than others. It also had a particular transport problem in that when the main steam railways had been built it had been impossible to bring trains into the central area. As a result, all the termini were in a ring around the area where people really wanted to go. There was simply nowhere to put trains in the crowded and narrow streets.

There was only one way to go, and that was to bury them. In the 1860s the first "underground" railways opened in an attempt to solve this problem. Large trenches were dug out, and where it was necessary to restore the surface they were bridged over. This was the so called "cut-and-cover" method. The Metropolitan railway managed to link the northern stations, Paddington, Euston, Kings Cross, while near the river the confusingly named Metropolitan District railway served Westminster.

There was a rather severe problem with this. The only means of propelling the trains was by steam engines, which would be traveling mostly in tunnels. Despite the engines being fitted with devices to absorb the steam, the conditions for the passengers were foul and smoke-filled. The Metropolitan's owners even had the nerve to claim that the "invigorating" atmosphere "provided a sort of health resort for people who suffered from asthma", but they also allowed drivers to grow beards in a futile bid to filter out the worst of the fumes. A civil servant who had spent time in Sudan said the smell reminded him of a "crocodile's breath". One attempt to improve conditions saw smoking banned, until an MP objected and insisted that all railways provided a smoking carriage.[20]

Clearly something better was needed, and in 1890 the first section of true underground railway was opened, of course using electric traction. This was dug deeper, minimizing the disturbance on the surface, but only clean and fume-free electricity was suitable. This section now forms part of the Northern Line. Though the largely windowless trains were not popular (there is nothing to see in a tunnel, is there?) it had shown the way.

In 1897, Boston opened its first section of underground railway (or subway). This was meant to solve a problem in a congested area of the center and the solution was to run the streetcars underground for a section. They stuck to the cut-and-cover method and obviously used the same overhead conductor used by the streetcars. As the cars came up into the streets for the rest of the route this was the only practical solution. With the construction method the height of the tunnels didn't matter too much.

Back in London, the Waterloo and City Line opened in 1898 and then the big step forward 2 years later was the beginnings of the Central Line from Shepherds Bush to Bank, going through the center of London following the line of Oxford Street and its extensions. This introduced an innovation; it was all one class and the fare was 4 cents regardless of how far was traveled. As the stretch was about 6 miles this could mean as little as a two-thirds of a cent per mile.

It was at this point that Frank Sprague re-enters the story. The early underground trains had used an electric locomotive to pull the train, which was a natural extension of railway practice. Particularly on the Central Line this gave problems as the vibration annoyed property owners above the line. It had been realized for some time that a system of motors underneath the carriages, as the streetcars used, was the way forward. The difficulty was to control all of these from the driver's cab. Sprague's idea, which he took from elevator control, was to run only low-voltage control wires along the train and have each carriage pick up its own power. This way, the difficult and dangerous high-voltage cables didn't need to be run along the train.

Sprague instaled such a system in an elevated railway in Chicago South Side in 1898, and it was so successful that it rapidly spread. In 1901, the Central Line converted to the Sprague multiple car control system.[21] It is now standard for underground and suburban trains. In passing, Sprague made another contribution, designing the original "dead man's handle" spring device that ensured the driver was in a fit state to control the train.

After all the worries about monopolies built up by railway barons, and hence the restrictions placed on the tramway operators, it seems extraordinary that an American robber baron was allowed to get control of much of the London underground system. In 1898, after he was forced to leave Chicago, Yerkes came to London, buying an interest in the struggling District Railway.

He turned his attention to the as yet unbuilt deep-level tube, the Charing Cross Euston & Hampstead Railway company (CCE & HR, now part of the Northern Line). He obtained Parliamentary authorization to build the line and became chairman of the company. By March 1901, Yerkes had control of the District Line. Forming the Metropolitan District Electric Traction Company (MDET) in July of that year, Yerkes raised £1 million of capital, mostly from America, to invest in the company.

Next, the MDET acquired control of the Brompton and Piccadilly line as well as the CCE&HR in September 1901. The third subway line Yerkes had in his sights was the half-finished Baker Street & Waterloo (now known as the Bakerloo) line. It was acquired by the MDET in March 1902. After raising yet more money this became Underground Electric Railways of London (UERL), which later became known as the Underground Group (Fig. 4.4).[22]

Thus it was Yerkes' drive and financial acumen that got the early subway lines built, and ensured the electrification of the Metropolitan and District systems in 1905/1906. These had linked together to produce the Circle Line but worked in an uneasy alliance. The company went on to success, with sections of the Bakerloo and Piccadilly Lines opening in 1906. Yerkes didn't live to see this; he died in 1905 with his personal finances in a mess.

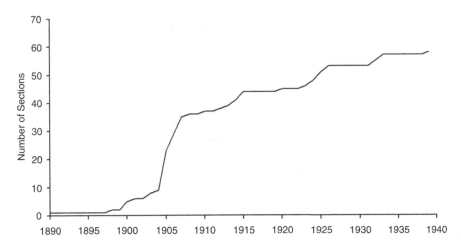

Fig. 4.4 Growth of the London underground, showing the opening of sections of the system, including the electrification of the Metropolitan and District Lines, and a measure of the growth of the system. Source: Author.[23]

The next city to open an underground railway or subway was New York in 1904. In a sense, they had lost the race with Boston by many years, but what they built was a true subway system which was much larger at 9.1 miles and 26 stations. As the line was deeper they used the same system as in London with power supplied through the rails. Its advantage was that the tunnel could be made a little smaller.

In all the cities, most of this initial work was within the existing built-up area but, particularly after World War I, the subway system started to expand into the countryside and this was when it started to have its impact, particularly on the growth of London. This was despite the fact that many of these places had been served by steam-hauled mainline trains for many years. Golders Green, which Yerkes had been keen to reach, was a greenfield site until the arrival of the Northern Line in 1907 triggered a building boom. This reflected the "where the rails go, the houses will follow" philosophy of the UERL.[24]

A spectacular example was the station at Rayners Lane on the western outskirts of London, beyond Wembley and Harrow. In 1929, 30,000 passengers a year used the service there which, though it sounds a lot, is only 80 people per day. A decade later, on the eve of World War II, the numbers had jumped to four million a year or 11,000 a day, reflecting the vast increase in housing in the area.[25]

It was clearly in the interests of the underground railway companies to encourage this process as it brought more custom for their railway systems. Perhaps the biggest exponent of this was the Metropolitan Railway up to 1933 before it, like all of the streetcars, buses and subways in the London area, was absorbed by the London Passenger Transport Board, commonly known as London Transport.

Each year from 1915 to 1932 the Metropolitan published a guidebook to new housing being built along their lines, entitled *Metroland* and priced at 4 cents. The descriptions of the villages which lay between the outskirts of London and the small towns beyond were supposed to be enticing, but sound faintly sinister now. Pinner was described as "just a sleepy little village living a secluded life of its own" until the coming of the Metropolitan Railway. Chorleywood: "Beautiful woods, a glorious open common, and a salubrious atmosphere have attracted many new comers to the favored locality". Harrow's "residential advantages are specially advanced by the admirable train services. The electric trains are rarely at rest".[26]

To the south of London, where the underground barely penetrated, the railway companies started to electrify their routes. They found that the investment paid for itself in the reduced operating costs (steam locomotives need pre- and post-operative attention and expensive maintenance) and that the improved acceleration and timekeeping made for a more efficient service.

At first the lines were those that started in London and went a short way out, but gradually the electrification extended, eventually reaching the south coast. Again there was the rise of such suburbs as Coulsdon, Purley and Epsom, Other railway companies followed suit, and the Euston to Watford Junction section and surrounding lines were electrified, as was the Liverpool Street to Southend line. By the 1930s nearly every suburban railway in the London area was electrified. The electric train had clearly shown its superiority for the short sections and frequent stops on these services. The result was that Greater London's population grew from 7.5 million in 1921 to 8.7 million in 1939. The move to the suburbs was even greater than these figures suggest as the numbers in the inner city fell by almost 450,000 over the same period.[27]

Other advanced cities around the world began the same process of electrifying suburban services, once again aiding the movement of population to outlying areas. Thus much of the building in cities was in these neighborhoods. For example, in the interwar period some four million houses were built in the whole of Britain, with 90% in new or existing suburbs.[28] It was the fast, suburban electric transport that made this possible.

The trams and streetcars lasted until around World War II in Britain and America, when they began to be a problem with the growing road traffic. The loading points in the centers of the roads were decidedly unsafe, and when the trams stopped, they held up all the traffic. They were replaced by trolley buses which could use a modified version of the overhead wires (the streetcars had used an earth return through the rails, but the trolley buses needed to provide this with a second overhead wire) but didn't need the tracks which were expensive to maintain. By the mid- to late-1950s even they were being superseded by motor transport.

Many European cities of middling size hung on to their streetcars, often because they were not really large enough to start going underground. In Britain the fashion to get rid of the trams probably went too far. Recently, they (or light rail systems) have been making a comeback with new systems operating in several major cities which use a combination of street running in the city centers and, where available, former conventional rail lines in some suburbs.

The electric trains, both underground and surface, are largely still with us. The cities couldn't function without them. With mainline services it is a different story. In America, few outside the Northeast Corridor have been electrified. With the long distances and relatively low usage the economics favor simply replacing the locomotives with diesels.

Right across Europe, steam was replaced by electrification, though in Britain this didn't happen until the 1950s and 1960s. This was only partial and the other lines used diesels. Most of these were in reality electric locomotives with a diesel-powered generator on board. The wheels were actually driven by electric motors. The reason for this was the ease with which control could be achieved. Gradually other sections, such as the East Coast mainline have been electrified as the message that speed pays has sunk in. The decline in passenger numbers of the 1960s and 1970s, due to the rise in car ownership, has been arrested and from the 1990s has shown a sharp rise.

The exploitation of electric power has had an enormous effect on public transport. It revolutionized the acceptability and fares of lines that ran in and out of towns and cities and allowed them to expand and breathe more easily. This was largely what made possible the suburbs of the towns and cities that are home for such a large part of the population.

NOTES

1. Robbins M, The early years of electric traction. *The Journal of Transport History,* 21:1, 92.
2. Sprague F.J. The solution of municipal rapid transit. *American Institute of Electrical Engineers Transactions,* 5, 352–393, September 1887 to October 1888, reprinted *Proceedings of the IEEE,* 72:2, February 1984.
3. NNDB, Frank J. Sprague, available at: http://www.nndb.com/people/904/000173385/.
4. Harvey & Press, p. 14.

5. *Bristol Mercury*, May 13, 1898.
6. Palmén M, Bristol Tramways Power Stations 1895–1941. Southwestern Electricity Historical Society supplement no. 30 to Histelec newsletter.
7. Hannah L. *Electricity before Nationalisation*, p. 16.
8. Byatt I.C.R. The British Electrical Industry, 1875–1914, p. 29.
9. Data for US are from: Wikipedia 'List of streetcar systems in the United States' though this understates the position as some systems have no start dates and have hence been ignored; Data for UK are from Wikipedia 'List of town tramway systems in the United Kingdom' which bases its information on K. Turner, *Directory of British Tramways* and other more local sources.
10. Semsel, C.R. More than an ocean apart: Street railways of Birmingham and Cleveland. *The Journal of Transport History*, 22:1.
11. Hattersley R. *The Edwardians*, p. 434. One British pound was worth $4.87 in around 1900 and as there were 12d (pence) in a shilling and 20 shillings to the pound, there were 240d to the pound. Thus 1d was about 2 old cents.
12. Hartley, p. 25.
13. Schrag Z.M., Urban mass transit in the United States, Columbia University, available at: http://eh.net/encyclopedia/urban-mass-transit-in-the-united-states/.
14. Barratt P. Chicago's Public Transportation Policy, 1900–1940S, available at: http://www.lib.niu.edu/2001/iht810125.html.
15. Hepp J. Streetcars, The Encyclopedia of Greater Philadelphia, available at: http://philadelphiaencyclopedia.org/archive/streetcars/.
16. Boston Streetcars, Streetcar suburbs of Boston, available at: http://www.bostonstreetcars.com/streetcar-suburbs-of-boston.html.
17. This can be seen by comparing maps from the 1890s with the 1913 tramway map in Harvey & Press.
18. Comparing tramway maps in Hartley with growth map in H. Clout, *The Times History of London*, p. 88.
19. Hartley, p. 25.
20. Watts P. London underground's history, available at: http://www.timeout.com/london/big-smoke/features/2814/London_Underground-s_history.html.
21. Robbins, p. 98.
22. London Transport Museum, Information Resources, Charles Tyson Yerkes, available at: http://www.ltmcollection.org/resources/index.html?IXglossary=Charles+Tyson+Yerkes.
23. It is difficult to get a true picture of the growth of the Underground with the varied developments on numerous lines. This diagram uses the opening of the various sections of lines as a crude measure. The data are based on Clive's Underground Line Guides, available at: http://www.davros.org/rail/culg/ and Douglas Rose's mapping of the changes. Retrieved from http://www.dougrose.co.uk/.
24. Hartley, p. 36.
25. Dimbleby D. How We Built Britain, p. 233.
26. Dimbleby, p. 233.
27. Clout, p. 112.
28. Hunt T. *Building Jerusalem*, p. 448.

5

First You Have to Make It:
The Spread of the Electricity Supply

The day must come when electricity will be for everyone, as the waters of the rivers and the wind of heaven. It should not merely be supplied, but lavished, that men may use it at their will, as the air they breathe. In towns it will flow as the very blood of society. Every home will tap abundant power, heat and light like drawing water from a spring. And at night it will light another sun in the dark sky, putting out the stars. There will be no more winter, summer will be eternal, warmth will return to the old world, melting even the highest snow.

Emile Zola, *Travail*, Book III, Ch II, pub. 1901

At the beginning of 1900 there were 170 electricity supply undertakings in Britain; 100 of them were municipal organizations and 70 private companies. By the end of 1907 this had more than doubled to 389 with 233 municipal and 156 private, which was two-thirds of the number that were nationalized in 1948. Around the turn of the century was the great time for setting up to supply electricity (Fig. 5.1). America had a much smaller proportion of municipal organizations. In 1902 there were 2805 private and 815 municipal; in 1907, 3462 private and 1252 municipal, which rose to 4224 private ones in 1917 before starting to decrease.[1]

What characterized these organizations, particularly in Britain, was their small size. In 1907, less than a quarter had more than a 1000 customers and only 2 had more than 10,000. At the other end, more than a quarter had less than 300 customers with 26 of them supplying under a 100.[2] There were 33 undertakings in London alone. The whole set-up was a shambles.

In America, the widespread adoption of street railways or trams, unimpeded by obstructive legislation, had given the electrical industry a head start. It was to give the country a lead which was maintained for most of the twentieth century. Not only had expertise in generation been built up, but it was also easy for these companies to sell surplus power for local lighting schemes. Most of these were still small local systems. However, there were organizations with the ambition to supply larger areas, for example in Westchester, where the Westchester Lighting Company had a plan to supply 11 surrounding towns.[3] In New York, the Consolidated Gas Company, finding that its gas lighting business was

© Springer International Publishing AG 2018
J.B. Williams, *The Electric Century*, Springer Praxis Books, DOI 10.1007/978-3-319-51155-9_5

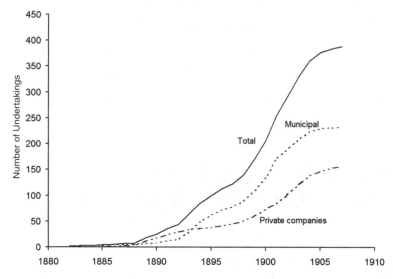

Fig. 5.1 The golden age for setting up electricity undertakings in Britain, dominated by municipal organizations. Source: Author.[4]

declining, took over some 12 electrical supply companies and had ambitions to build a large local generating station.[5]

In Britain, one of the contributory factors to the small scale of the organizations was the legislation. There was a need for some, as the gas companies had the powers to dig up roads to lay their pipes but electricity undertakings could not do this without seeking a specific Act of Parliament. This left them at a considerable disadvantage, and no real competition could take place until this was straightened out.

In 1882, Joseph Chamberlain was the President of the Board of Trade, and hence responsible for electric lighting. In the 1870s, while he had been Mayor of Birmingham, he had been very involved in what was known as the "Municipal Gospel", which was that local authorities should control the monopoly provisions such as water and gas. The profits from these could be used for development and for the benefit of the people of the town or city.

It was natural that he should want to extend this to electric lighting, and also be concerned that it might impact on the substantial profits made from the municipal gas undertakings. Hence the 1882 Electric Lighting Act was heavily biased towards municipalities, which had to give their consent to a private company setting up in their area, and in any case could take over their assets after 21 years. It seemed to be assumed that ultimately the industry would be in public ownership.

Unsurprisingly, there were few takers. The ones that did were private companies taking a chance, though most were taken over by their municipality later. The local authorities were more cautious with ratepayers' money at stake, and were as much concerned by the state of the technological development as anything else. They were probably wise to delay.

With a change of Liberal to Conservative government, in 1888 another Electric Lighting Act was passed.[6] This was a little friendlier to private enterprise, raising the time before the local authority could take over to 42 years and improving the terms. It was also made easier to obtain orders, and the Board of Trade could overrule local objections if it saw fit.

This, together with improvements in the equipment available, was sufficient for the industry to begin to take off. However, there were still the problem of the conflicting claims of local councils who wanted to control what happened in their area versus the need to generate centrally and supply as large an area as possible. In addition, the technical issues of what sort of supply it should be could not be resolved.

Electricity can be divided into two types: direct or continuous current (DC) where the current always flows in one direction, and alternating current (AC) where it increases in the forward direction reaching a peak, decreases and then flows in the opposite direction, reaching a peak there before subsiding again. This is one cycle, and the number of these per second gives its frequency, which is measured in Hertz (Hz).

As usual when the answer is not clear cut, the debate over which was best generated a lot of argument but little consensus. For lighting, it generally didn't matter which was used. Motors really needed to be on DC as, until the work of Nicola Tesla in the 1890s, there was no practical motor that would run on AC. Even then the AC motor runs basically at a speed determined by the frequency of the supply, whereas that of the DC motor can be made variable. Certainly in the early stages DC had an advantage here, and so all streetcars or trams and trains ran on DC.

As was seen in Chap. 3, the electricity supply could be subject to large peaks of demand. On an AC system, the generators had to be capable of supplying that or customers would have to be switched off. With a DC system, rechargeable batteries could be used to assist in supplying the peak demand. In reality, they could not be very large and this was only useful on small-scale systems, and shows how limited were the ambitions of the supply undertakings.

It was in the size of the system that the critical factor came into play. Even good copper wires have resistance, and current passing along them causes a "voltage drop", meaning that the voltage at the customer's premises is somewhat lower than that being generated. As more customers use the system and the distances from the generator become larger this becomes a bigger and bigger problem. Eventually it becomes unacceptable, which was what many of the Metropolitan Electric Supply Company's customers were complaining about.

With a DC system, it was difficult to do anything about this, so only customers within a mile to a mile and a half of the generating station could be supplied. At the beginning this was about the sum of most of the supply companies' aspirations, but the limitations soon became apparent. In sizable cities, it meant that large numbers of generators would need to be dotted about.

With an AC system there was a simple solution: the transformer. A coil of wire with a changing current passing through it generates a magnetic field. If a second coil is in the same magnetic field, for example on the same core of iron, then a current can flow in this secondary circuit. It is easy to have a different number of turns in the coil, which produces a different voltage. It is a delightfully simple and robust device.

AC can more easily be generated at a higher voltage or transformed up to it. As the power delivered to the customers is voltage times current, the required current is much lower, and hence the losses in transmission are much reduced. Near the customer the voltage can be transformed down again to a suitable level. It seems like magic, but the problem is largely solved.

The DC protagonists tried to emulate this system, but the only way to step the voltage up and down was by using a motor and generator set at each point. It is one thing to have a transformer at each substation where the high voltage is brought down to the level needed by the consumers, but quite another to have a piece of rotating machinery which requires supervision and maintenance.

The leading scientists and engineers took sides, each claiming the primacy of their system, with almost messianic fervor. In America, Edison was for DC and Tesla for AC, and hence General Electric was for DC and Westinghouse for AC. There the battle became one of commercial and industrial systems and often threatened to turn nasty.[7] In Britain it was conducted in a more gentlemanly manner, but just as dogmatically.

The result was that each small-scale operation (which they were because of parochialism and sometimes the bias of the legislation), could choose its own type and voltage, depending on the whims of its chief engineer and which camp he belonged to. The result was, predictably, chaos. Adjacent systems were usually incompatible and so there was no hope of connecting them together to produce larger networks.

In America, ruthless commercialism helped to rationalize some of the systems, while in Germany public/private cooperation produced larger undertakings for whole cities. In London, parochialism reigned supreme. For example, in around 1911 Chicago and Berlin each had six power stations, while London (admittedly some three times the size) had 64. The sales of electricity per head of population in those cities was 83 kWh for Berlin, 291 for Chicago, and only 49 for London.[8]

Clearly London, and Britain as a whole, was falling behind in the electrical race. This could be seen in the total amount of electricity used in the three countries. In 1905, Germany used one and a half times and the United States four times as much as Britain.[9] In succeeding years the gap closed to some extent with the Germans, and increasingly with the Americans. In 1908, the incoming president of the Institution of Electrical Engineers, W.H. Mordey, tried to show that Britain was not really behind by comparing various towns and cities around the world, but it merely pointed to the enormous variations and that any conclusion was possible.[10] He did point out that the generation capacity was less in Germany than in Britain, but as the output was more this merely highlighted that the load factor was better, being an average of nearly 16% in Germany and an appalling 11% in Britain.[11]

The country that had prided itself on leading the industrial revolution was starting to fall behind in this growing new field. It was as much in failing to grasp these opportunities as in the decline of the staple industries of textiles, shipbuilding and coal that Britain lost its supremacy.

What was becoming clear was that the greater the size of the power station, the more efficient it was, so the electricity could be sold at a lower price, hence increasing the number of consumers. A virtuous circle resulted. This was not possible in most places in Britain because of the local parochialism embedded in the legislation. While this had not seemed unreasonable in 1882, when the concern was largely to prevent private monopolies, and the successful municipal takeovers of gas and water undertakings stood as a model, its later effect was very damaging. There was always this dichotomy in electricity supply between the greater efficiency of large systems and the worries about monopolies. The mess had to be sorted out later, and that became expensive.

There were, however, some bright spots. Of all the people who contributed to the development, four stood out. The first is included, not because he was successful at the time, but because he pointed the way. The second is there because he made the essential breakthrough

Fig. 5.2 Sebastian Ziani de Ferranti. Source: https://en.wikipedia.org/wiki/Sebastian_Ziani_de_Ferranti#/media/File:S_z_de_ferranti.gif

that led to the large-scale low-cost generation of power, and the third and fourth because they demonstrated that large-scale networks were possible, by building them.

By 1890, the system of the London Electricity Supply Company (LESCo) stretched from Regents Park to the River Thames, and from Knightsbridge to the Law Courts and had some 312 customers.[12] It was an AC system which generated and transmitted at (as they said at the time) a "pressure" of 1200 V, and had a transformer to step this down to 100 V in each consumer's house.[13] They were still using wires over the roofs.

The considerable difficulties they had run into led them to turn to a young man called Sebastian Ziani de Ferranti, and appoint him chief engineer (Fig. 5.2). Despite his name he was born in Liverpool, but his photographer father was an immigrant who could trace his lineage back to Doges of Venice.[14] Even though he was young, he had also built up considerable experience in the infant technology.

Ferranti solved their immediate problems, but thought he could see where the future lay. There were considerable difficulties with generating power locally, not least in that the noise and vibration annoyed the neighbors. The solution, he felt, was to move the power station to Deptford where noise would no longer be a problem, and they could access cheap coal brought in by water and use the river for cooling. A large station could be built which would transmit power economically to central London by laying cables alongside the tracks of the railway companies. To minimize the losses he proposed the unprecedented "pressure" of 10,000 V.

You had to admire his courage, or was it foolhardiness? Surprisingly, the company backed his plan, and he set to work to build a generating station bigger than any other at the time. The generators were so large that they were driven by 10,000 hp engines and had to be designed by Ferranti and built on site. That was not all; he then needed to invent a paper insulation suitable for the transmission cable to central London.

Ferranti was technically brilliant, but he had overreached himself and the company only just survived as he couldn't make it all work. Time was to show that Ferranti was basically right, but he was too far ahead of the current technology, and it was some years before all the problems of large-scale generation were overcome. When they were, the systems looked extremely similar to what he had been trying to build.

Ferranti went on to build a large manufacturing business. It became a major supplier of transformers and his inventive genius led to many advances in the technology, but it was only when his wilder ideas were restrained and firm financial controls were implemented, that the company became successful.[15] He was a great advocate of large-scale systems, and was to play a role later in the development of the grid.

The trend to larger and larger generating sets had exposed the limitations of reciprocating steam engines. One of the biggest problems was that they only rotated at something like 500 rpm which required indirect connection via pulleys and ropes to the generators which needed anything from 1500 to 4800 rpm (Fig. 5.3). The search was on for a better prime mover than the machines in use which were often derived from railway locomotive designs.

Charles Algernon Parsons is best known for racing his small turbine-powered ship, "Turbinia", around the fleet at the Spithead Naval Review in 1897. It seemed like a prank, but it had the serious purpose of demonstrating the potential of the steam turbine for propulsion to the Navy. It worked, because a few years later "Dreadnought" and later battleships were all turbine powered. However, Parsons' contribution to electric power generation was probably even greater.

Fig. 5.3 An early electricity power station. The steam engines on the *left*, which look suspiciously like railway locomotives, are coupled with ropes to the small generators on the *right*. Source: A. Byers, *The Willing Servants*.

He started with considerable advantages in life, but made careful use of them. As the youngest of six sons of the third Earl of Rosse, a keen amateur astronomer, he received a good education from scientifically-educated tutors and his interest in mechanical matters was encouraged. He then went to Trinity College Dublin for 2 years before going to St. John's College Cambridge to study mechanism, applied mechanics and mathematics.[16] Cambridge, at the time, still didn't have an engineering department.

Not content with a merely theoretical education he began his engineering training with a 4-year apprenticeship at the works of Sir William Armstrong & Co. at Elswick, Tyneside. This was followed by 2 years with Kitson & Co. of Leeds, where he developed a four-cylinder high-speed epicycloidal steam engine that he had patented, before obtaining a junior partnership in the firm of Clarke, Chapman & Co. of Gateshead, in 1884.[17] As head of the newly-organized electrical department, he began to consider the problems of designing high-speed generators and an engine to drive them.

It had been known since Heron of Alexandria (c. CE 62) built his "Sphere of Aeolus"[18] that power could be obtained from jets of escaping steam. Later work had shown that if the velocity of steam expanding from a high pressure could be harnessed, more energy could be extracted than was possible in a conventional reciprocating steam engine. The problem was that the speed it achieved when escaping from the pressurized nozzle to atmosphere was far too high to be practical.

Charles Parsons' stroke of genius was to realize that, by dividing the expansion of steam into a number of pressure drops, the speed at each stage would be reduced, and hence the blades of a turbine would rotate at a practical level. He took out his first patents in 1884, and they show how thoroughly he considered all the problems to be overcome in making such a high-speed turbine.

The same year he built his first turbo-dynamo, solving not only the problems of the turbine, but also the high-speed dynamo. The device produced 7.5 kW at 18,000 rpm. It was an immediate success and in the next 4 years about 200 were made, mostly for ship lighting. He showed that not only could he invent but could turn the ideas into practical devices.

In 1889, the company at last found someone with the courage to put turbines into a generating station. The Newcastle and District Electric Light Co. Ltd. was set up to do this but they took the precaution of making Charles Parsons the managing director. Clarke Chapman built two 75 kW turbo-alternators and they were installed in their new generating station at Forth Banks, Newcastle.[19]

During that year Parsons' frustrations with Clarke Chapman boiled over, and he dissolved his partnership and founded his own company, C.A. Parsons & Co. at Heaton, about two miles from the center of Newcastle. Unfortunately, he wasn't able to recover his patents from Clarke Chapman, so he invented another form of the turbine where the steam went radially instead of parallel, which avoided his own patents.

By January 1890, Forth Banks was commissioned, making it the first turbine-powered generating station. Parsons understood the need to demonstrate the performance of his machines so he managed to get the commission to set up an electric lighting scheme in Cambridge. The Cambridge Electric Supply Co. was set up with Charles Parsons as managing director, and installed three radial flow 100 kW units which were successfully brought into service in 1892. Tests showed that their efficiency was equal to the best reciprocating engines of similar power.

The next year he brought two 120 kW units into operation in Scarborough, and then in 1894 the Municipality of Portsmouth installed a 150 kW set, but ran it with reciprocating sets alongside; not a great vote of confidence. It was then that he had a stroke of luck. The Metropolitan Electric Supply Company was in trouble. (Yes, them again, though this was before their difficulties in Chap. 2.) Their Manchester Street generating station was the subject of complaints from the neighbors about the vibration from the reciprocating engines. This had gone on for some years without a successful remedy being found. Eventually, in 1894 an injunction was granted against the company.

In desperation, MESC approached Charles Parsons, and he agreed to build a 350 kW turbo-alternator, more than double the power of his previous machines, in the hope that, as it could be run without being bolted down, this would solve the problem. It did this successfully, demonstrating that it could save the station, and more similar sets were rapidly installed. These were the parallel flow type as Parsons had managed to recover his patents. He was now in a position to really demonstrate the full capabilities of his system. Westinghouse in America was convinced and acquired the U.S. patent rights.

The next big step was a pair of 1 MW generators for the city of Elberfeld in Germany, whose successful operation convinced the Swiss company Brown Boveri to also take up a licence. The sizes steadily increased: 1.5 MW by 1901, then 3.5 MW by 1905, 6 MW in 1908, and a huge jump to 25 MW in 1912 for the Fisk Street Station in Chicago. Unlike the conventional steam engine, the design was readily "scalable" and larger and larger units could be made relatively easily.

The success of the turbine for electricity generation is shown by its percentage of the installed capacity, from around 5% in 1901–1903, to 45% in 1905–1907 when the patents had run out and all the main manufacturers piled in, to 75% in 1908–1910 and 80% in 1911–1913.[20] Later, this was to reach virtually 100% and the size of the sets 500 MW and beyond. It has carried the industry ever since, with only the gas turbine, itself a derivative, making any inroads (Fig. 5.4).

Fig. 5.4 A Parsons 2 MW turbo-alternator set made for Newcastle Electric Supply Company. The large cylinder center *left* is the generator and to its *right* is the turbine, which is a similar size. Source: Parsons et al., "Steam turbines", IEEE Journal, 1904, 33:167, 794–809.

Just as remarkable was the increase in efficiency that Charles Parsons achieved during this period. The efficiency was measured by how many pounds weight (lb) of steam was needed to produce a kW of electricity. This decreased from 27.9 in the Scarborough station, to 18.2 in Elberfeld, to 10.45 in the Chicago Fisk Street Station. Even that was beaten in the Carville station near Newcastle, when it almost reached 10 lb/kW in their 11 MW sets.[21]

What Charles Parsons had achieved was to free electricity generation from clanking steam engines, and give it a future where giant generators would be capable of supplying the ever-increasing demand. It had taken 20 years of concentrated effort and faith before the turbine really took off, but as well as his ability to invent and improve, he had the personal and business skills to see it all through.

This achievement alone would rank him with the greats, but he also revolutionized marine propulsion, not only in the Navy but also in the great age of liners such as the "Lusitania" and "Mauretania". The "Mauretania" held the Blue Riband of the Atlantic for the fastest crossing to America for nearly a quarter of a century.[22] Not content with that Parsons became involved with building large telescopes. Altogether he took out more than 300 patents, but unlike so many inventors he had a calm and methodical approach and so was able to be successful in almost everything he touched.

Someone else with these attributes was Charles Hesterman Merz; he is a name unknown outside the electrical engineering fraternity, but he also made a huge contribution to the development of electricity supply (Fig. 5.5). Like Ferranti he was the son of an immigrant, Theodore, who had married into the Quaker business community of Newcastle, and with his own chemical company was a leading industrialist in the area. He was also a very cultured man and is known to scholars as the author of *A History of European Thought in the Nineteenth Century*.[23]

Fig. 5.5 Charles Merz. Source: R.A.S. Hennessey, *The Electric Revolution*.

Instead of going to university, in 1892 Charles Merz became a trainee at the Newcastle Electric Supply Company (NESCo), which had been set up 3 years earlier by his father and an uncle. After 2 years, he continued his apprenticeship at the Robey engineering works in Lincoln. This enabled him to get a position with British Thomson Houston (BTH) superintending contracts, which soon led to him being made engineer and manager running a power station in Croydon.

The next step was to superintend electrical installations at a new station in Cork, Ireland. It was here that he met two men who were to have an important role in his life, William McLellan and R.P. Sloan. By now he had gained so much experience in power stations that he was consulted by another uncle, Wigham Richardson, about a plan of the Walker and Wallsend Gas Company to also provide an electricity supply. He was critical of the scheme being proposed and was soon appointed to replace the existing consultant, which meant that he presented the engineering case when their Bill went before Parliament.

As he was successful, the company offered him the post of Chief Engineer, but he declined because he wanted to set up as an independent consultant where he would have greater scope. Aided by McLellan and Sloan, he started to design a power station at Neptune Bank near Newcastle. The most important decision was to use AC so that a wide area could be served, and three phase which was far better for rugged industrial motors. This was quite courageous as the AC induction motor had only been introduced a few years before, but he could see that the future lay not just in providing lighting, but serving as wide a range of customers as possible.

He next had to choose a frequency. This had to be higher than the 25 Hz used by some companies for industrial power, as lights flickered on it, and lower than the 83 or 100 Hz used by some companies for lighting as this was too high for the motors. He compromised on 40 Hz which would serve both types of customer. This was his one mistake, though there was no way of knowing it was wrong at the time; 30 years later they had the expensive process of converting to what was by then the standard of 50 Hz.

By the time the station opened in 1901, the company and NESCo had realized that it was foolish to compete, and the station was taken over by NESCo to serve both companies' customers. So successful were NESCo at attracting customers, which wasn't difficult given the business and family links to major companies in the area, that almost immediately it was necessary to plan an extension.

An important year for NESCo was 1902, as they were part of a consortium to promote a Bill for an electric tramway on the north bank of the Tyne. One of the objectors was the North Eastern Railway which was concerned that it would lose traffic. During the proceedings Merz persuaded the Chairman of the Railway, George Gibb, that the solution was railway electrification. This resulted in a deal for NESCo to supply the railway with power and make common use of some cables and substations.

To supply this increased load, the output of Neptune Bank was more than doubled from 2.8 to 5.8 MW by installing a pair of Parsons 1.5 MW turbo-alternators. Merz had become convinced that turbines were the correct way forward. He was now in a position to bring all his ideas together in the design of the next NESCo station, Carville. When it opened in 1904, it was equipped with two 2 MW turbo-alternators, soon followed by two 3.5 MW sets.[24]

Merz' skill could be seen in that Carville was the lowest-cost power station in the country, and its running costs were as low as the best. With his partner he wrote a paper which was for many years regarded as the definitive work on power station design.[25] This established his reputation.

It wasn't just in the technical matters, but also in politics and in the commercial field that the Merz clan showed their ability. They were quite prepared to repeatedly go to Parliament to increase their franchise area. One of these applications gave the company authority to supply most of the Northumberland coalfield. By 1906, electricity was cheap enough there for the houses of some of the miners in Ashington to have electric light.[26]

They were particularly successful in their commercial developments. The approach was to absorb or work with as many of the surrounding supply organizations as possible. In 1905, the company purchased the two County of Durham electric power companies, whose owner was in financial difficulties, and immediately connected them to their system, standardizing the voltages and frequency. When, further south, the Cleveland and Durham County Electric Power Co. was also in difficulties it decided to buy bulk power from NESCo's low-cost Carville station. Gradually, most of the independent companies in the area took their power from NESCo as their network expanded. Ultimately, NESCo was to take control of most of the companies.

Before World War I NESCo was supplying most of North East England, but despite having much higher costs the Newcastle and District Company and the Corporations of Darlington, South Shields, Sunderland and West Hartlepool remained separate; a clear case where parochialism triumphed over commercial logic.

To build a network this big, two problems had to be solved. One was to keep the transmission losses to an acceptable level, which Charles Merz solved by raising the transmission voltage from 6 kV to the unprecedented level of 20 kV. The other matter was to protect the network when a fault occurred. He realized that this had to be automatic and he and a colleague introduced the Merz–Price system of protection.

The achievement of Charles Merz and his colleagues was to build the biggest integrated power network in Europe by 1912.[27] It stretched from the coalfields of Northumberland to Cleveland south of the Tees. They produced 12% of the county's electricity, but with 6½% of the plant, meaning that their load factor was 24½% whereas the rest of the country only managed about half that.[28] This was the secret of their success: highly efficient generation, low capital cost, a wide range of customers which gave a high load factor. The result was that NESCo was producing 60 times the amount of electricity they had in 1901, compared with less than 7 times for the rest of the country. They could sell it much cheaper, and still be profitable.

Even in 1905 the way forward was clear enough for Charles Merz to propose a scheme for sorting out the situation in London. The proposal was to build three large power stations along the river and supply low-cost power, particularly to industrial customers who were not being well served by the existing suppliers. It was necessary to get a Bill through parliament for this, but despite his skill in cleverly modifying his Bill to get support, it became mired in politics and doomed by a change of government.

However, Merz was not finished with this subject. He had the outstanding success of NESCo as his model. Where some would have retreated, he cautiously went forward. He was now thinking much bigger, with the whole country as one giant three-phase AC network, with a small number of large, highly-efficient power stations; but that was still some way in the future.

Samuel Insull was also an Englishman, but he had gone to America and such was his business acumen that he had become the secretary to Thomas Edison. However, in 1892 he went Chicago, a town that had more than 20 companies producing electricity, and became the president of the small Chicago Edison company, one of many Edison franchises around the country.[29]

He quickly understood the essentials of the business and the first was to improve the "load factor" by changing the type of customers so that power could be supplied more evenly throughout the day. He then went on a remarkably similar path to Charles Merz and NESCo. The choice was to generate AC so that it could be transmitted larger distances. To get more efficient generation he turned to turbines, installing one of the first in America in the Fisk Street station.

With an inheritance of DC stations in the center of the city he pioneered the use of rotary converters. These were basically a pair of coupled machines a DC motor and an AC generator or vice versa. Thus, for example, the customers who were supplied with DC could continue to have it even when the inefficient local generating station was shut down.

The next step was to exploit the economies of scale and that meant taking over as many of the local power companies as possible and connecting their customers to his efficient generating stations. By 1907, he had acquired 20 utility companies and renamed the company Commonwealth Edison. Like NESCo, he had built a very successful company on a very similar set of rules, and it stood out as one of the most progressive in the world.

Insull had created a virtual monopoly in Chicago, and this didn't make him very popular despite falling electricity prices. He went on to create a holding company which owned ever more utilities. However, the stock market crash of 1929 exposed the dangers of this financial manipulation and the whole castle came tumbling down. The various companies went bankrupt one by one and so did he personally.[30]

He fled to Europe but after skipping from country to country he was eventually extradited back to America. Here he was acquitted of all charges and was then able to quietly retire to Paris. He had set the utilities on the right path, but like others before him, he had overreached himself; this time it wasn't technical, but in business.

NOTES

1. Hausman W.J. and Neufeld J.L. The Economics of Electricity Networks and the Evolution of the U.S. Electric Utility Industry, 1882–1935, presentation to the 2004 Annual Meeting of the Business History Conference, Le Creusot, France; Quinquenial US special censuses of the electrical light and power industry 1902–1907, available at: http://eh.net/database/quinquennial-u-s-special-census-of-the-electrical-light-and-power-industry-1902-1937/.

2. Figures are derived from tables in J.D. Poulter, *An Early History of Electricity Supply*, pp. 190–205. Companies that were subsequently taken over by municipalities have been classified as municipal. The number of companies at nationalization from L. Hannah, *Engineers, Managers and Politicians*, p. 7.

3. *New York Times*, December 2, 1900.

4. Data processed from Poulter, pp. 190–205.
5. *New York Times*, February 10, 1900.
6. See Electricity Lighting Acts of 1882 and 1888.
7. Hennessey R.A.S. *The Electric Revolution*, p. 71.
8. Hughes T.P. *Networks of Power*, p. 258.
9. Hughes, p. 4; British figures from L. Hannah *Electricity before Nationalisation*, p. 427.
10. Mordey W.H., Inaugural Address, *Journal of the Institution of Electrical Engineers*, 42:193, 10–22, Feb 1908.
11. Load factors calculated from: German installed capacity from Mordey, Output from Hughes, and British figures from Hannah.
12. Hannah, p. 11.
13. Electricity Council, Electricity Supply in the UK: A chronology c. 1987 entry for 1885.
14. DNB, Ferranti, Sebastian Ziani de (1864–1930).
15. DNB, Ferranti.
16. DNB, Parsons, Sir Charles Algernon (1854–1931).
17. Hannah, p. 13.
18. Derry and Williams, *A Short History of Technology*, p. 313.
19. Parsons R.H., *The Early Days of The Power Station Industry*, p. 171.
20. Figures derived from Table 21 in Byatt, p. 112.
21. Parsons, pp. 178–183.
22. DNB, Parsons, Sir Charles Algernon (1854–1931).
23. This whole section is derived from four sources: DNB, Merz, Charles Hesterman (1874–1940), Hannah, pp. 28–33, Byatt, pp. 11–122, and Hughes, pp. 249 and 451–457.
24. Parsons, Stoney, and Martin, The steam turbine as applied to electrical engineering. *Journal of the Institution of Electrical Engineers,* 33:167, 794–809, 1904. There is some discrepancy in the sources about the size of the smaller sets, but presumably Parsons knew what he was supplying.
25. Merz and McLellan, Power station design, *Journal of the Institution of Electrical Engineers,* 33:167, 696–742, 1904.
26. Arthur M. *Lost Voices of the Edwardians*, pp. 218, 224.
27. Hannah, p. 33. They were in a race with Rheinisch Westfälische Electricizitäts AG (RWE) in the Ruhr valley, but at this stage were still ahead.
28. The figures are computed from Byatt, p. 114 and Hannah, pp. 427–432. It will be noted that they don't agree with those often quoted, e.g., Hannah, p. 33, which don't agree with calculations from his own figures.
29. Emergence of Electrical Utilities in America, available at: http://americanhistory.si.edu/powering/past/h1main.htm
30. Samuell Insull (1859–1938), available at: http://www.chicago-l.org/figures/insull/

6

Beginnings of Mass Production:
Electric Power in Industry

I shall never forget Mr. Boulton's expression to me: "I sell here, Sir, what all the world desires to have—Power."

James Boswell, 1774

At first sight it seems counter-intuitive to use a steam engine to drive a generator and then to turn the electricity back into mechanical power with an electric motor. However, there turned out to be numerous advantages, sufficient for it to take off at great speed after about 1900, particularly between 1901 and 1905 when the prices of electric motors halved[1] at the same time as their reliably becoming far more satisfactory.

Figure 6.1 considerably understates the true position in America as the majority of the industrial and railway supplies in the early years were generated by the organizations themselves. In 1907 and 1912 some two-thirds of the industrial electric power was being generated privately, though the proportion was slowly falling.[2] Unsurprisingly, the industrial load from public supplies follows the increase in the number of supply organizations, but with a lag. Curiously, the industrial demand exceeded the others a little earlier in Britain than in the US.[3]

Despite a slow start, from around 1900 onwards the changeover to electric motors across all industry was relentless. It averaged 15–20% per decade up until the start of World War II, staying well ahead of the percentage of homes connected to the mains supply. Though some sectors were very conservative in their approach, the rapid adoption in others made up for this. It is interesting to note that there was barely a pause during the First World War. The heavy use of electricity in the munitions manufacture compensated for a slowing elsewhere (Fig. 6.2).

In Victorian times the typical arrangement for supplying power to machines in factories was to have a central boiler and steam engine which drove line-shafts. These were long rods that usually ran the whole length of the building mounted below the ceiling. On it were pulleys from which flat belts some two or three inches wide came down to the individual machines. No factory inspector today would pause for a moment before condemning such a dangerous arrangement (Fig. 6.3).

© Springer International Publishing AG 2018

J.B. Williams, *The Electric Century*, Springer Praxis Books, DOI 10.1007/978-3-319-51155-9_6

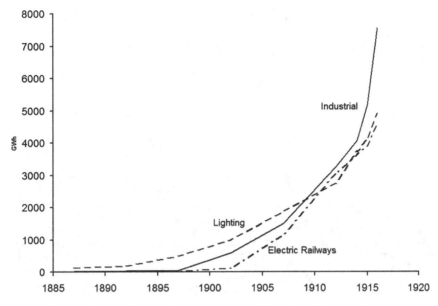

Fig. 6.1 Growth of the public electricity supply in the US. Industrial use rapidly overtook the use for lighting and for electric railways. Source: Author.[4]

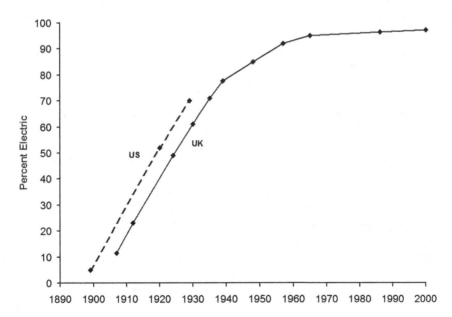

Fig. 6.2 Growth of electric power usage in industry. Source: Author.[5]

Fig. 6.3 Overhead line shafting driving a machine shop. The pulleys on the shafts can be seen close to the ceiling with the belts coming down to the individual machines. Source: http://www.lathes.co.uk/southbend/page11.html.

There was no simpler way of subdividing the power of a single steam engine and distributing it to drive individual machines. To stop a machine the belt, while still moving, had to be pulled over on to an idler pulley. A similar process was needed to change the speed of the machine. Obviously, the working machines would have to be arranged to suit the line-shaft and not put where would be most convenient for the operation that they were doing.

Once electric motors were installed to drive the line-shafting, and watt meters were attached, the extraordinary inefficiency of this method was discovered. The losses were measured when just the line-shafts were being powered, with no machines connected. The best that were found absorbed 22% of the power when loaded, with 50% being quite normal.[6] Some older systems lost over 80% of the power.[7] In an ideal world, individual motors were attached to machines, and power was distributed by electric cables, but there was considerable conservatism and investment in existing plant that held back adoption.

One of the most profitable places to introduce electric power was where it was replacing manual labor. The Vickers shipbuilding and armaments company, for example, were strong early users of electricity, employing 1300 motors across their various works by 1903.[8] Their 40-ton Siemens furnaces had traditionally been loaded, or as they called it "charged", by hand. This required a team of four highly-paid men and took 4 h. It was a very heavy job and it was noted that this time could be longer in hot weather. It was found that it could much better be done by an electrically-driven Wellman charging machine, and it only took half an hour.

The savings were not only that it freed men, requiring only two instead of four, but it released them from the heavy work to concentrate on the skilled part of their task. The real advantage was that 25% more output was obtained from the furnaces. Also for the cost of a small amount of electricity the wages of two men were saved which could be used elsewhere.

Vickers had another example where a gang of six men was needed to raise the heavy doors of a large furnace. It was found that this could easily be achieved with a 1½ HP motor, freeing the men to do something more useful. The advantage came from the ability to distribute the power around to awkward places in a large works, or where the required position was remote from the central engine. Another example was electricity used to drive hoists, where great savings could be found in replacing strenuous labor.[9]

Coal mining was an area where there should have been opportunities, but the coal industry was notorious for its attitude to innovation. In the UK, NESCo found that the Tyneside engineers naturally came to them to discuss their power requirements, but the Durham coal owners had to be offered inducements to even talk to them.[10] It was easy to power the major items from steam engines on the surface as there was an abundance of very cheap, low-grade waste coal from the mine, but underground was a different matter.

Attempts had been made to use compressed air underground but these had not been very successful and around the turn of the century electricity was being introduced for coal cutting, carrying the coal from the face to the shaft, and air and water pumping. It is not surprising that mine managers were reluctant to proceed until proper standards for equipment were introduced and flameproof motors were available. Though enclosing motors to prevent the propagation of a flame was proposed in 1905 it wasn't until 1921 that the Safety in Mines Research Board was set up and testing began in the following year, though some manufacturers had their own testing regimes long before this.[11]

Despite the difficulties some progress was made, though the bulk of electricity underground was used for pumping and hauling the coal, and not in cutting it. The only real advantage for using electric coal cutters came on narrow seams, otherwise miners were cheap and plentiful. In 1902, there were 145 machine coal cutters but by 1912 this had risen to somewhere between 1200 and 2000.[12] Put another way, 2% of coal was machine cut in 1900, 8½% in 1913 and by 1938 this had risen to around 60%, with wide variations between coalfields depending on their geology and the conservatism of their management.[13] For mining as a whole, 4% of the power in 1907 was supplied by electricity, but this had risen to 40% by 1924. Gradually electrical power was being used to replace some of the more backbreaking work underground and improve conditions for miners.

Shipyards were quite different and took to electric power easily. In Britain, some of this might have been because around half of them were in the North East and so could have been persuaded by Charles Merz and NESCo.[14] The scattered layout of the yards and the need for power in many different places made line-shafting very difficult and a scattering of small steam engines was very inefficient. Electrification was started with cranes where mobility was needed and power could be picked up with a tramcar type arrangement, but moved on to the winches, punching and shearing machines, plate rollers and drilling machines around the yard.[15]

The shipbuilders soon found that the flexibility of electric power was much more adaptable and within a few years they were using far more power than they had when they were dependant on steam.[16] At the start they had been concerned about the relative costs, but that soon disappeared and they began to realize that they could achieve greater output, which was the important factor.

Table 6.1 Degree of electrification in various industries in the 1920s

	Great Britain (%)	Mean of US and Germany (%)
Coal	41	48
Iron and Steel	42	72
Rolling mills	50	78
Cotton	20	60
Engineering	91	97
Shipbuilding	92	87
Blast furnace	30	47
Paper	54	56
Dyeing	53	72
Non-ferrous metals	67	92
Electrical Engineering	100	99
Wool	96	60

Source: Author[17]

The flexibility of electrical motors could be seen in the clothing industry. This had been revolutionized in the last part of the nineteenth century by the introduction of the oscillating shuttle sewing machine which could sew more than just in straight lines. The uptake was such that by 1907 94% of firms were using machines.[18] For factory production, turning these machines by hand or treadle was tiring, and by the same date 60% were powered which was much more efficient.

In the 1890s, electric motors started to appear, and rapidly gained so much ground that in 1907, 57% of the powered machines were electrically driven.[19] The motor could either by fastened under the table where the treadle would normally go, or just behind the stem of the machine coupled with a simple belt to the rear of the hand wheel. By 1924, 80% of the machines were powered and of those 84% were electrically driven, reaching 90% by 1930.[20]

This set the stage for the mass production of clothing. Despite the numbers of sewing machines in the home (Singer alone sold some three million machines in Britain between 1880 and 1905 and sales in America ran at well over a million a year around the later date[21]) the manufacturers had a distinct advantage as electric motors for home machines only really started to appear after World War I. By then the acceptability of ready-made clothes was already a reality, and the clothes shop rather than the dressmaker became the usual place to go. The improved quality of the off-the-peg garments meant that they were barely distinguishable from the bespoke items worn by the rich. It became more difficult to categorize people instantly by the clothes they wore, bringing a greater equality.

It might have been expected that there would have been a considerable uptake of electrical power in the textile industry, but it wasn't the case; in 1907 it had only reached 5% in Britain though it was a bit more advanced in the US.[22] The reasons are not hard to find. The textile factories had put a lot of effort and investment into having efficient steam engines, and though they were using line-shafts the regularity and constant running of the machines meant that the system was reasonably satisfactory. It most cases it didn't matter that weaving produced better cloth at the more constant speed which could be achieved with AC motors.[23] It only made sense to switch to electric power when a new factory was being built, and as in this period the industry was slowly beginning to decline, little new plant was being installed. Despite this, electricity was powering about a quarter of the industry in 1924.[24]

Where the advantages of electric motors really brought benefits was in the industrial machine shop. Here the lathes, milling machines, shapers and planers were of all different sizes and needed to be started and stopped at different times. A single machine could be run without having to run the whole line-shaft. Also, for example in the case of a break-down, the whole works didn't have to stop.

Machines could more easily be run at varying speeds by simply turning a lever without the complex business of shifting the belt from one pulley to another on a moving line-shaft. This was so time-consuming that often the operator didn't bother, or reckoned that the loss by running at the wrong speed was less than the time lost to change the speed. With the electric motor the speed could be quickly changed and the machine run at its optimum speed, giving a subtle increase in production.

These advantages were sufficient for rapid adoption in the industry. The whole category of engineering, shipbuilding and vehicles had reached 42% powered by electricity in 1907, rising to 67% in 1912, and 88% in 1924.[25]

The flexibility brought further changes as it was realized that it wasn't necessary to group machines of the same type together as had been common practice, but they could be arranged where they were convenient for the flow of work. A work piece could then be simply moved from machine to machine as the various operations progressed until it reached the end as a completed unit.

In the nineteenth century, the rules for "tolerances" had been formulated into a system known as "limits and fits". When an item is machined it is never the exact size, but as long as it is sufficiently close to the desired result then it is acceptable as being "in tolerance". Where two parts have to fit together the tolerances can be arranged so that if one is as big as it is allowed to be and the other as small, they will still fit together. If the parts are at the other end of their tolerances, though they may be slightly loose, they will still work satis-factorily. This is the basis of "interchangeability of parts" as it is commonly known.

The place where everything came together was in the fledgling car industry and particu-larly with one man: Henry Ford. He had grown up on a farm near Dearborn, Michigan, not far from Detroit, and had shown an interest in mechanical things from an early age. In 1896, he built his first automobile, the Quadricycle, but he had the ambition to go into auto manufacture. In 1903, he set up the Ford Motor Company, but it only produced a few vehicles a day and these were assembled by groups of men bringing parts to a single place and gradually building the motor.

Over the next few years the company created a number of reasonably successful mod-els, but Ford really wanted to build an automobile for everyone, and in 1908 he introduced the model T, or Tin Lizzie. It was as simple and strong as he could make it, but he knew that price was the key to the size of the market, and that was the way he wanted to go.

As sales of the model T rose, in 1910 the company moved to a new, and much larger, factory at Highland Park, and with the move began the reorganization of the method of production to keep pace. One of Henry Ford's skills was to surround himself with compe-tent men. It was here that his team of production engineers set to work. While it can be argued that neither Ford nor his team actually invented anything, their key ability was to gradually put together small improvements to create a highly-efficient system.

They started by organizing the machines so that the work flowed from one procedure to the next. To do this required electricity, and lots of it. They had installed a 3000 HP gas engine, giving some 2 MW of power.[26] This was greater than the capacity of more than three-quarters of the power companies in Britain at the time, and sufficient to supply a small town.[27] Although small sections of line-shaft were used, these were driven by electric motors. That was the only way to get the flexibility needed to speed up the flow of work.

Next the team tackled the assembly processes, starting with the magneto. On April 1, 1913 the assemblers found themselves next to a waist-high slide carrying the units instead of the workbenches at which they normally assembled the whole unit. Now they were required to put together a number of parts and then slide the work along for the next man to add his section. This was no April Fool joke; it raised the production rate from one magneto every 20 min per man to one every 13.[28] They knew they were on the right track.

There were difficulties with some workers being faster than others, but this was simply solved by having a powered chain conveyor to move the work pieces, which set the pace of the line. The team tinkered with the height of the line and other details, gradually over the following year removing people until the production time fell to five man minutes per item—a staggering improvement.

Rapidly the team moved on to other subassemblies such as engines and transmissions, and converted these to moving lines, bringing immediate increases in productivity. Then they tackled the big one: the chassis assembly. This went through a considerable development process until the balance and electric-powered chain speeds were got right, but by April 1914 the man hours to assemble the main chassis had dropped from 12½ to just over 1½, a very marked improvement stemming from the ease with which parts could now be supplied to a fixed place, and the elimination of people walking about.

Ford had managed to produce a virtuous circle. In 1908, the price of the Touring car version of the model T was $850 and 5986 had been sold, but in 1914 the price came down to $490 and 260,720 vehicles were sold.[29] Though the price had been pushed down in the intervening years, it was the introduction of assembly line methods that brought the large jump. There was, however, a problem. The workers hated it, and the labor turnover was horrendous. But it didn't take Ford long to find a solution. The labor rate was more than doubled to $5 a day and conditions improved. The turnover dropped sharply and the whole thing settled down. Despite the reductions in price, efficiency had risen sufficiently to absorb these increases.

Henry Ford was quite clear what he was doing: "Every time I reduce the charge for our car by one dollar, I get a thousand new buyers," he said.[30] Step by step he and his team had created the assembly line. They had put together bits of ideas from many different sources, but what they finally produced was revolutionary. Without the electric power to drive everything from the individual machines to the assembly lines and feed conveyors, it wouldn't have been possible. The appetite for power can be seen from the installation of more than another 3 MW of capacity.[31]

If Ford had been secretive about what he was doing it wouldn't have had such an impact. He was quite open; he could afford to be, he was well ahead. There wasn't just one secret, it was the system, the attention to detail, that brought the advantages. As a result of his openness, the assembly line spread rapidly through manufacturers in the United States. It migrated from cars into domestic appliances and radios, and on from there, giving American industry an advantage for some time to come.

It was, however, a Faustian bargain. Everyone could have cheap and plentiful goods, but the price was that some people would have to work as part of the machine that produced them. They were generally well paid for this, but it was not an easy accommodation. The car plants had a long history of industrial strife which lasted a good deal of the century. Now, with much of the more mundane or unpleasant work being done by robots, they are more peaceful. Elsewhere, with better industrial relations, the production lines were more successful.

Ford, with the help of his electric power, had laid the foundations of a consumer society.

NOTES

1. Byatt I.C.R. *The British Electrical Industry, 1875–1914*, pp. 88–89.
2. Calculated from Table 17 in Byatt, p. 95. Only these two dates are given as they correspond with censuses of production.
3. UK information from Hannah L. *Electricity before Nationalisation*, pp. 427–428.
4. Data from Electrical World, 80, 546, 1922.
5. UK: 1907, 1924, 1930, from Censuses of Production. 1912 from Byatt, p. 75, further points by scaling the increase in electricity by its increase in consumption. 1935, 1939, 1948 from Hannah, and others from Department of Energy and Climate Change: Historical electricity data: 1920 to 2010; US: Smiley G, The US economy in the 1920s, available at: https://eh.net/encyclopedia/the-u-s-economy-in-the-1920s/; Parson E. The 1920s (1920–1929), available at: http://ecmweb.com/content/1920s-1920-1929
6. Byatt, p. 88.
7. Calculated from figures for case 1 and 2 in Chatwood A.B. Electric driving in machine shops. *IEEE Journal,* 32:163, 964–983, 1903.
8. Williamson A.D. Applications of electricity in engineering and shipbuilding works. *IEEE Journal,* 32:163, 925–964, 1903.
9. Wraith W.O. Description of the electrical equipment of an engine works and shipyard, with notes thereon. *IEEE Journal,* 33:166, 994–1015, 1904.
10. Byatt, p. 94.
11. A History of Mine Safety Research in Great Britain. Part of a Health and Safety Executive Report (1986). Available from: http://www.users.zetnet.co.uk/mmartin/fifepits/starter/safe-2.htm
12. Byatt, p. 93.
13. Greasley D. The diffusion of machine cutting in the British coal industry, 1902–1938. *Explorations in Economic History,* 19, 246–268, 1982.
14. At the beginning of WWI. In T.P. Hughes, *Networks of Power,* p. 446.
15. Anderson J.A. The distribution of electricity in shipyards and engine works. *IEEE Journal,* 33:167, 84–856, and also Wraith.
16. Hughes, p. 456.
17. Data from: Edgcumbe, T.D. President's Inaugural Address, *Journal of the Institution of Electrical Engineers, 67*:384, 1–11, 1928.
18. 1907 Census of Production.
19. 1907 Census of Production.
20. 1924 and 1930 Censuses of Production.
21. Godley A. The development of the UK clothing industry, 1850–1950: Output and productivity growth. *Business History,* 37:4, 46–63.
22. Byatt, p. 74.

23. Whitehead S.N.C.K. Individual electric drive in modern weaving sheds. *IEEE Journal*, 51:222, 860–867, 1913.
24. Byatt, p. 76.
25. Byatt, pp. 74–76.
26. Batchelor R. *Henry Ford*, p. 41.
27. Calculated from the data in J.D. Poulter, *An Early History of Electricity Supply*, pp. 190–207.
28. Hounshell D.A. *From the American System to Mass Production, 1800–1932*, p. 248.
29. Hounshell, p. 224.
30. Nevins A. *Ford*, p. 493.
31. Hounshell, p. 229.

7

Early Mass Media: Newspapers and Cinema

The cinema is an invention without a future.

Louis Lumière

In printing, particularly of newspapers, it was the controllability of electric motors which was the particular advantage. The papers had shown steady growth in their circulations in the nineteenth century, which was partly due to more universal education, but also (in Britain at any rate) the removal of the stamp tax. Just as important, if not more so, was the use of the railways to distribute the papers in time for breakfast and, crucially, the ability to print large numbers very quickly.

This is particularly acute for a daily morning paper, where the latest possible news needs to be included. Thus the moment when the journalist's prose is finally locked up is late into the evening. This has then to be typeset, though the introduction of the Linotype machines in the 1890s greatly speeded this up, as the text was entered on a keyboard and the output of the machine was a cast line of type. Though the early machines were driven by lineshafts, they were a natural for electric drive and were soon converted as suitable motors and supplies became available.

The beds of type then went through a process known as stereotyping where a mold was made of the page when was then used as a form to make the actual plates for the press. This had the advantage that multiple copies could be produced and also they could be curved to fit the drums of the rotary presses. Once these processes were complete somewhere between 5 and 10 h were left to print the papers. To sell a million papers required printing at between 100,000 and 200,000 per hour. If you couldn't print at this rate, no amount of clever journalism, or gimmicks, would achieve the sales.

Since Robert Hoe had perfected the rotary press in 1844 in New York, newspaper presses had gone through considerable development by the company he formed. Once the basic arrangement was mastered they had the clever idea of duplicating the sections. By 1887 they had produced a Quadruple machine capable of printing 48,000 eight-page papers per hour. Four years later, their Sextuple machine could print the same number per hour, but now of 12 pages or 72,000 of 8 pages. By 1902, progress was such that the Double Octuple press could produce 104,000 16-page papers per hour.[1]

© Springer International Publishing AG 2018
J.B. Williams, *The Electric Century*, Springer Praxis Books, DOI 10.1007/978-3-319-51155-9_7

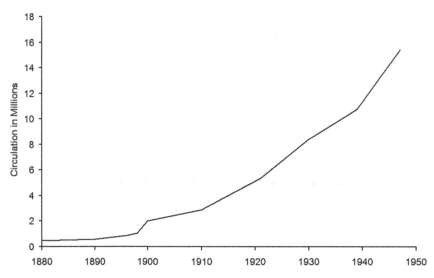

Fig. 7.1 Growth of circulation of the main daily newspapers in Britain, 1880–1950. Source: Author.[2]

At first sight the sharp increase in circulation of daily papers in Britain in 1900 would seem to have been caused by the introduction of electric driving, but it isn't so (Fig. 7.1). The introduction of the *Daily Mail* in 1896 and the *Daily Express* in 1900, combined with the thirst for news of the Boer War, is quite sufficient to explain the rise. Of course they had to have the latest presses to achieve the numbers, but though those of the *Express* were electrically driven, the *Mail*'s were too early and were only converted later.

Though experiments had taken place earlier, it was only from about 1898 onwards that satisfactory electrical systems began to be installed. The particular problem was that the press needed to be able to be "inched" in very small steps while being set up, and then steadily run up to its full speed so that the roll of paper did not break. With steam driving of line-shafts this meant manipulating belts on pulleys; a skilled and very dodgy process.

A simple electrical system consisted of two motors, the smaller one heavily geared down to provide the "inching" and to start the press running up. This was then disconnected by a clutch and a main motor took over to power it to the normal working rate. Improvements soon followed where the whole press could be controlled from a small panel with push buttons for the "inch" and running it up to speed. Thus these monsters were brought under control, which could only be achieved with electrical systems.

By 1901, some 50 newspapers in Britain had adopted electric driving either completely or in part.[3] Between 1903 and 1906, *Lloyd's Weekly* paper, which had a circulation well over a million, tried a number of competing approaches on its seven Hoe Double Octuple presses (Fig. 7.2). Electric systems were taking over and it was only a matter of seeing which method was best. By 1907, more than 50% of all the power used in the printing of newspapers and magazines had turned over to electricity.[4] Five years later this was 84%, rising to 94% in 1924.[5]

In America, with its much greater distances, a national press was more difficult and the very large numbers of papers were largely local.[6] However, at the end of the nineteenth century a considerable amount of consolidation took place at the hands of people such as

ONE OF SEVEN DOUBLE OCTUPLE PRESSES USED FOR LLOYD'S WEEKLY NEWS

PRINTING & FOLDING COMPLETE UP TO 32 PAGES.

Fig. 7.2 The massive size of rotary printing presses. Note the man standing by the ladder and the space underneath for the electric motors. Source: G.A. Isaacs, *The Story of the Newspaper Printing Press.*

Joseph Pulitzer and William Randolph Hearst, increasing the print runs for the survivors. They too could invest in the mammoth presses.

The way was open for a mass press. After the end of World War I the numbers sold just kept rising. In Britain they were equivalent to half the households in the country, and by World War II to all of them. Thus newspaper reading, which had been a minority interest in the nineteenth century, had become something for everyone, with a much better informed populations even if some of the what they were reading was not of the highest quality!

*

On February 20, 1896 Robert William Paul screened a film called "Rough Sea at Dover" and some others at Finsbury Technical College in London, using the "Theatrograph" projector of his own design. While he had a few technical problems, these were soon ironed out and he patented his improved projector at the beginning of March.[7]

He was rapidly hired by enterprising businessmen to hold regular showings at venues around London. First he appeared at the Egyptian Hall, Piccadilly, from March 19, and then from for 2 weeks at the Alhambra Theatre of Varieties in Leicester Square. This was so successful that he remained there for 2 years and was kept so busy that he spent his evenings traveling from music hall to music hall rewinding the films on the way.

Very soon, Paul produced his own camera and began making films. To show his expertise he filmed the finish of the Epsom Derby in June, showing the Prince of Wales' horse "Persimmon" winning. He processed the film overnight and screened it to an enthusiastic Alhambra audience the next day. This was one of the first news films.

Paul was really a maker of electrical instruments, and had got into the film business in a rather strange way. He had been approached by two Greek businessmen to make copies of Edison's Kinetoscope, which the great man had omitted to patent in Britain.[8] This was a moving picture device which could only be viewed by one person, like a peep show. When viewed with hindsight, this seems a rather strange direction for development to take off, and was an example of where it headed up a blind alley. Edison had thought he could make more money from selling multiple devices.

After completing the job for the Greeks, Paul made more to sell himself.[9] Understandably it was difficult to obtain films for these, so Paul collaborated with American photographer Bert Acres to design a camera and produced some films. They quickly fell out; Paul felt betrayed when Acres patented the camera solely in his own name, so he carried on alone.

It was a short step, but a critical one, to displaying the Kinetoscope films on a screen where many people could watch them at the same time. His objective was to bring the illusion of motion to the magic lantern. To do this he had to make a projector. The critical device he came up with was the Maltese cross or Geneva mechanism for rapidly advancing the film frame by frame. This could be powered either by a hand crank or, as Paul saw, by an electric motor (Fig. 7.3).

It also needed a powerful light source. His approach was to adapt the lamp house from a magic lantern which was used to display still photographs. While various lamps had originally been used these rapidly settled down to three main systems. The most powerful was the arc lamp where an electricity supply was available. Next came the limelight, where a block of lime was heated to incandescence in an oxy-hydrogen flame. The gases were supplied from cylinders and connected by tubes to a burner which heated the lime.

Fig. 7.3 Robert Paul's Theatrograph projector. The handle on the large wheel is for winding the film through. The big *black* box at the rear is the lamp house. It isn't clear which type of light is being used. Source: http://cinemathequefroncaise.com/Chapter3-1/Figure_03_06_Theatrograph.html

The third method, only suitable for small venues, was to use an acetylene flame with the gas being made by a generator where water was dripped on to calcium carbide.

Sales of his projectors and cameras increased rapidly and they were widely used for the next 10 years. By the turn of the century they were being exported to the continent as well as Australia and other British dependencies.[10] This was in addition to his original business of electrical instruments.

Paul also had a third strand to his activities: film making. He began with "actualities", small slices of real life, but soon realized the potential of short comedies such as "A Soldier's Courtship" filmed on the roof of the Alhambra Theatre where he was exhibiting. This starred Fred Storey, Julie Seale, and his wife Ellen as the interloper.[11] It became so popular that he made a second version the following year.

He realized the potential of drama and comedy, and in 1898 set up the first studio in England, at Muswell Hill in London, complete with a laboratory to process the film. During that summer he produced over 80 short dramatic films.[12] He also covered a wide range of subject matter including topical events such as the Boer War in South Africa. He entirely dominated the home market and it is no wonder that he earned his title "Father of the British Film Industry".[13] However, by 1910 he had become disillusioned by show business and pulled out, burning his stock of negatives and selling his film-making equipment.

Like all discussions on who was the first at anything, it is never that simple. In fact, on the same day Robert Paul made his first showing, the Lumière Brothers Cinématographe equipment was shown in London though it had been seen in France at the end of the previous year. The simultaneous appearance of both systems kick-started the cinema business in Britain. Soon many people piled in and the business exploded with a vast array of equipment and sizes and types of film.

William Dickson, though of British parentage, worked for Thomas Edison on the Kinetoscope, but later left, believing that the future lay in projecting films. He set up the American Mutoscope company right at the end of 1895. C. Francis Jenkins invented the Phantascope projector in 1893, which was improved by Thomas Arnat. However, Edison soon realized his mistake and, unable to produce a satisfactory projector, bought the Phantascope, renamed it Vitascope and sold it as his own invention. On April 23, 1896 the first showing of a film using this was shown at Koster and Bial's Music Hall in New York City.[14] Soon the system was being used around the country.

Another pioneer was the American Charles Urban, who first made his name with the Bioscope projector. It proved so successful and popular an invention that it became a generic term for cinema itself. Today it is still used in Holland, Belgium and South Africa. Urban had his first Bioscope projectors made in the United States as an improvement on the Edison devices. He accepted a job in England at the Maguire & Baucus London office, bringing his projectors with him. He rapidly changed the fortunes of the company and also changed its name to the more English-sounding Warwick Trading Company.[15]

The company prospered largely on the back of the good performance of the Bioscope projectors, but it also started making films. These were actualities, covering real places and events, which brought an increased interest in what were now known as Bioscopes. Later, Urban set up on his own as the Charles Urban Trading Company, still selling the projectors and other equipment but making large numbers of travel, topical and science films. All these were sold under the Urbanora brand name.

One of the earliest users of what were often called Bioscopes, though they could equally well be using other equipments such as Theatrographs or Vitascopes, were the traveling showmen. Their theaters had elaborately painted fronts to attract the visitors, behind which was the auditorium where the films were shown. Power came from dynamos mounted on the traction engines to light the myriad lamps on the façade, and for the projectors. The whole show fitted into wagons for moving from place to place.[16]

Theaters also took to the movies as an "act" in their varied programs which could include a play, a musical section, a comedian, a storyteller.[17] The film was thus regarded as just another form of entertainment to put on a bill. The churches also took to showing films in their halls. They were motivated partly by wishing to bring people into their establishments, which might lead to them taking an interest in the services, but also if they were watching a movie show the men weren't in the bar drinking. The films were seen as a form of entertainment to which the whole family could go together.

Major events began to attract the cameramen. Charles Urban, for example, sent four cameramen to South Africa to cover the Boer War, and others from rival companies were also there. They produced short scenes which ran for a minute or so. These could be spliced together to produce longer pieces, but the pure novelty drew the audiences. Films of Queen Victoria's Diamond Jubilee and funeral were also popular.

This gradually gave way to the idea of a regular showing of "news" events leading to the "newsreel". In Britain, the first was Pathé's Animated Gazette, which appeared during the first week of June 1910, though Pathé had produced a similar arrangement in France a year or so before. Each week, a film was shown that was a compilation of short pieces showing topical matters. A whole series of imitators rapidly followed suit, and the idea spread around the world with the French companies of Pathé and Gaumont in the lead.[18]

While not as immediate as a newspaper, films introduced whole new audiences to information which they wouldn't otherwise have had. They had come to the cinema to be entertained, but they were also being informed along the way. Once World War I commenced films were an important source of information about how the conflict was progressing, so much so that governments produced their own propaganda newsreels. Between the wars was the golden age of the newsreel with them being issued twice weekly and becoming an important source of news for the regular cinema-going public.

Everything was going well but, as Nicolas Hiley said: "If celluloid film base had been only a fraction more expensive to produce, or just a little more fragile, it would have rendered it impossible for traveling showmen and entertainers to adopt the new moving pictures".[19] Even then there was a fatal flaw with the base material, cellulose nitrate: it was closely related to gun cotton and was highly inflammable, and because the nitrate contained the source of oxygen it could even burn under water. As a result, a fire, once started, was very difficult to put out.

Responsible venues took precautions by putting the projectors in fireproof boxes, but still there were fires. That more conflagrations didn't occur with the combination of a flame for the light, and what can only be described as dangerous film passing through the gate a short distance away, was a testament to the good design of the projectors. Despite that, fatal incidents did occur and this led to calls for licencing and regulation of the business.[20]

In 1909, the British government legislated and the Cinematograph Act came into force in January 1910. It brought in a system of licencing by the local authorities and required the projector to be housed in a fire-resisting enclosure. The effect of this was to rapidly curtail the showing of films in the ramshackle collection of theaters, halls and so on, and move them into purpose-built cinemas. (As an aside, some local authorities took "inflammable film" to refer to the content as well as the film stock, and started to control what was shown. This led to the setting up of the Board of Film Censors.)

In America, in 1905 the first Nickelodeon appeared. This was a storefront theater which got its name from the five-cent charge. The idea soon took off and Nickelodeons spread around the country. These theaters usually had live vaudeville acts as well as showing the short films. Around 1910, they started to be replaced by larger modern theaters.

Purpose-built cinemas started to appear in Britain around 1908, with a number of claimants for the first spot, but the numbers only really took off once the Act came into force in 1910. From then on the rise was astonishing, and the halls and theaters quickly stopped showing films. In 5 years something like 5000 cinemas were built in Britain. The numbers have to be approached with some caution as there are no reliable figures for this period so this relies on various estimates, but all sources agree that something like this total was reached.

In America by the late 1920s the number of movie theaters had reached 20,000 showing a greater enthusiasm for this form of entertainment. Though the numbers were to drop to around 15,000 in the depths of the recession in 1935, audiences held up remarkably well. This was the age of Busby Berkley and other escapist movies which meant that viewers could get away from, at least temporarily, the worst of the depressed economic life outside.

In Britain, the first cinemas made a lot of money and a mad scramble ensued with investors piling in with money signs in front of their eyes. Inevitably provision outstripped the audience which grew from 4 million a week in 1911, to 7 million in 1914, only reaching 20 million a year or two later[21] when many of the cinemas were already struggling.

While the collapse in numbers might not be quite as severe as shown in Fig. 7.4, two other factors also came into play. When conscription was introduced in the middle of the war, the cinemas lost many of their staff, particularly experienced projectionists.

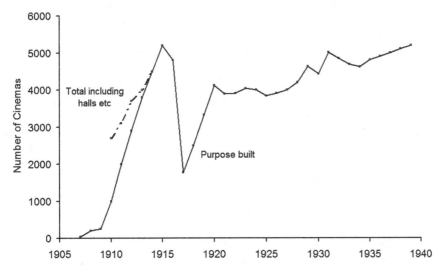

Fig. 7.4 Growth of cinemas in Britain, particularly purpose-built ones, 1905–1940. Source: Author.[22]

Also in May 1916 the government, needing funds for its war, introduced an Entertainment Tax which affected all cinema seats. It was quite severe, with the 2 cent (1 old pence) and 4 cent tickets attracting a 1 cent tax, and the common 4 cent to 12 cent range charged at 2 cents. This tax was increased the following year, aiming to raise a further 50% in revenue.[23]

Ironically, audiences either wanting more information about the war, or wanting entertainment away from it, remained high. The worries that the whole thing was just a craze and would go the same way as the roller skating rinks a few years earlier, eventually turned out to be unfounded, though perhaps the cinemas were saved from this by the war and their increasing maturity with better and longer films being shown.[24]

The names of many of the cinemas and the groups behind them pointed the way that things were progressing. Among them were Electric Theatres (1908) Ltd., Electric Palaces Ltd., London and Provincial Electric Theatres Ltd., United Electric Theatres Ltd., National Electric Theatres Ltd., Associated Electric Theatres Ltd., and Standard Electric Theatres Ltd. Even when not owned by one of these groups, the individual cinemas often had "Electric" as part of their name.[25]

With some sort of electricity supply available in most sizable towns and cities—and even if it wasn't it could be generated on site—it became an important part of the attraction and many establishments sported brilliant lamps or arc lights to make them stand out. This made the arc lamp the natural choice for the projector's lamp house. Also as the cinemas became bigger there was a need for brighter sources. If the distance to the screen is doubled, four times as much light is required for the same illumination on the screen. With arc lamps the current could be increased to obtain more light whereas the limelights had a fixed limit and so they began to die out.

The disadvantage of arc lamps for lighting was that the carbon rods required adjustment and replacing periodically, but the projection room was manned constantly and it was a simple task for the projectionist to attend to these. Some electrical equipment was required to reduce the voltage to the 55–75 V needed for the arc, such as transformers, or rotary converters to provide DC, but once installed there was no need to get supplies of cylinders of gas and so on.[26]

At first, many projectors were hand cranked, but as the number of longer films steadily increased as World War I approached, the advantages of electric motors became obvious.[27] In many cases it meant that only one projectionist was needed. While the first was constantly turning the handle, a second was needed to prepare the next film. Audiences soon became very intolerant of breaks in the performance.

The early films are often called "silent" as they didn't have a sound track on them, but that didn't mean that the cinema was quiet. Often live music was provided which might (or might not) be relevant to what was being seen, but there were also pianolas or machines which played records which we would call gramophones. A number of devices had been invented which "synchronized" the sound with the film such as Edison's Kinetophone, the Kinemopera, and Clarendon's "speaking pictures".[28]

These worked to a greater or lesser degree. Some mechanically linked the gramophone to the projector, while others were electrical where the speed of the gramophone motor was adjusted to keep it in time with the film. They all suffered from the problem that there was no really satisfactory way of amplifying the sound from the record, and they depended on horns (like the His Masters Voice logo) that projected from the gramophone. In a hall of any size they were just not loud enough.

The solution was found, with the coming of electronic amplification, well after World War I, but for a time there was a battle of systems between sound being recorded on separate discs and sound encoded on the film itself. The synchronization problems with a separate disc meant that eventually it fell out of use, and the Western Electric system became general. By the mid 1930s nearly all cinemas had converted to being able to show talkies as they were known, though, surprisingly, it had little effect on the size of audiences. Electricity had completely taken over.

While the early cinema was driven by novelty, and then by audiences wanting a refuge from the street or some privacy in the dark, gradually the content of the films became the important factor.[29] To get enough light, the films had always been made outdoors, but clearly that couldn't continue. Gradually studios were set up, but that required sufficient lighting for the cameras to work under indoor conditions. The obvious answer was again, arc lights.

In Britain, the studios were largely in London and the Home Counties, and one of the factors in their siting was an adequate supply of electricity, though some such as Sir Alexander Korda's Denham studios had their own generators capable of supplying 4.5 MW.[30] The result was that with brilliant lighting anything could be filmed inside in a studio if a suitable set was made. (It has been suggested that the film stars' fashion for wearing sunglasses came from the need to protect their eyes from the arc lamps.)

However, in America filmmakers were restricted by Thomas Edison's patents. Looking to get away from him and find somewhere where the sun usually shone, they settled on Hollywood in California.[31] There is some discussion about when the first film was made here but it was certainly in the 1910s. After that many studios were set up and the industry didn't look back. Hollywood came to dominate the film industry both in America and abroad.

Between the wars, the American movie industry was so successful that, despite the efforts of local filmmakers, the bulk of the film in most countries came from America and particularly Hollywood. In Britain, amongst other countries, this led to a quota system to encourage more local output which was only partially successful. The result was that audiences everywhere were exposed to a glamorized version of American culture. While this appalled the establishment, it did widen people's horizons and developed their taste for consumer goods.

The filmmakers invented all the techniques we take for granted, from trick photography to the "fiction" story. All the methods such as panning and zooming, cutting and editing were perfected so that art could take over from the mechanics of production. The feature film became an attraction that everyone just had to see.

With the coming of another war the audiences went up and up in most countries, reaching some 30 million a week in Britain. However, during the 1950s there began a long decline, matched almost exactly by the rise of television viewing. In America the decline was initially precipitate but leveled out after 1965. In Britain, it started more slowly but carried on down. Though there was some recovery from 1985 with the advent of the multiplex cinemas, the baton had been passed to the new contender (Fig. 7.5).

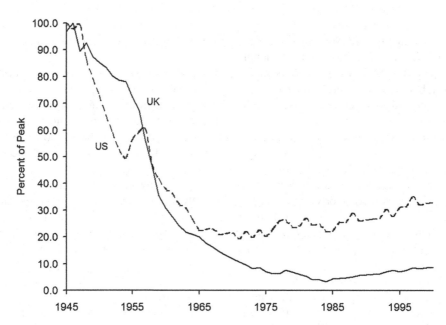

Fig. 7.5 Decline of cinema audiences, 1945–1995. Source Author.[32]

Everything wasn't lost as all the body of knowledge could be passed over. For a long time a good deal of television output was first filmed and only converted to electronic form for broadcast. Cinema had led the way and developed the market for mass entertainment and education beyond the wildest dreams of the theaters and music halls. In doing so, it broke down more social barriers by becoming something for everyone, and as they were all watching the same thing it was an important factor in making "one country". Without the electric cinema and electric studios it wouldn't have been possible.

NOTES

1. Issacs G.A. *The Story of the Newspaper Printing Press*, p. 81.
2. The data for this graph are obtained by adding up the circulations of all the main papers at the relevant dates. In some cases, this is not available so interpolations have been made from known information. The data come mostly from newspapers and from R. Williams, *The Long Revolution*, pp. 195–236, for the early part, though ABC data from 1921 onwards is used.
3. Issacs, pp. 244/5.
4. 1907 Census of Production.
5. 1924 Census of Production.
6. Stephens M. History of newspapers, *Colliers Encyclopedia*, available at: https://www.nyu.edu/classes/stephens/Collier%27s%20page.htm.
7. British Patent 4686 of 1896.
8. Croy H. *How Motion Pictures Are Made*, p. 34, also available at: http://www.archive.org/stream/cu31924026115554/cu31924026115554_djvu.txt.
9. DNB, Robert William Paul.

10. Barnes J. Robert William Paul, Who's Who of Victorian cinema, available at: http://www.victorian-cinema.net/paul.htm.
11. BFI Screen Online, Paul R.W., available at: http://www.screenonline.org.uk/people/id/449512/.
12. Pioneers, Robert W. Paul, available at: http://www.earlycinema.com/pioneers/paul_bio.html.
13. Barnes.
14. Dirks T. The history of film: The pre-1920s early cinematic origins and the infancy of film, available at: http://www.filmsite.org/pre20sintro2.html.
15. McKernan L. Charles Urban, motion picture pioneer, available at: http://www.charlesurban.com/bioscope.htm.
16. University of Sheffield, National Fairground Archive, available at: http://www.nfa.dept.shef.ac.uk/history/bioscopes/history.html.
17. E.g., The Palace Theatre, *The Times*, July 4, 1906.
18. Urbanora, 100 years of newsreels in Britain, *The Bioscope*, June 1, 2010, also available at: http://thebioscope.net/category/100-years-ago/. The magazine's commentator is still using Charles Urban's trade name.
19. Hiley N. At the picture palace: The British cinema audience, 1895–1920. In J. Fullerton (ed.) *Celebrating 1895: The Centenary of Cinema*, p. 96.
20. E.g. A fire in Newmarket, *The Times*, September 9, 1907.
21. Bottomore S. The coming of the cinema. *History Today,* 46:3, 14, 1996.
22. Data from multiple sources, generally following Hiley up to 1915, and L. Wood, *British Films 1927–1939* from 1927 onwards. In between, the figures are derived from the lists of theaters in the various Kinematograph Year Books. Where the data conflict a view has had to be taken as to the most likely figures. While numbers undoubtedly declined during World War I, the sharp drop in 1917 in the numbers registered with the KYB cannot be fully explained. The numbers being liquidated and wound up certainly exceeded 1000, but there still seems to be a gap. Perhaps in the confusion of wartime, they simply didn't register, but the steady recovery afterwards takes some explaining if the numbers weren't real.
23. *The Times*, May 10, 1916 and May 3, 1917.
24. Hiley N. Nothing more than a craze: Cinema building in Britain from 1909 to 1914. In A. Higson (ed.) *Young and Innocent: The Cinema in Britain, 1896–1930*.
25. Kinematograph Year Book 1914.
26. Bennett C.N. *Handbook of Kinematography*, p. 168.
27. Hiley, *At the Picture Palace*, p. 98, Fig. 3.
28. KYB 1914.
29. Hiley, *At the Picture Palace*.
30. Hennessey R.A.S. *The Electric Revolution*, p. 168.
31. History of Hollywood, California, available at: http://www.u-s-history.com/pages/h3871.html.
32. Derived from: UK: ONS Social Trends 40, Chapter 13 Lifestyles and social participation: Figure 13.8 Cinema admissions; US: Pautz M. The decline in average weekly cinema attendance: 1930–2000 adjusted for change in population.

8

The Catless Miaow: Wireless Telegraphy

The wireless telegraph is not difficult to understand. The ordinary telegraph is like a very long cat. You pull the tail in New York, and it miaows in Los Angeles. The wireless is the same, only without the cat.

Albert Einstein

The one area where the exploitation of electricity reached maturity in the Victorian era was the telegraph. This had begun commercially in the 1830s and by 1900 its cables bound the world together. Two years later, with the completion of a Pacific Ocean link, it encircled the world. The traffic volume was around 50,000 words a day.[1] The messages sent by the telegraph system were usually delivered stuck on to a printed form as a telegram. At the turn of the century the numbers of these reached a peak of around 100 million a year.[2]

For most ordinary people the telegraph system had no effect and it mattered little until the wars when telegrams were used extensively to pass on the bald news of a soldier's death to his family. After the wars there were attempts to expand its use; for a time greetings telegrams, for example when someone was not able to attend a wedding, were popular, but it never really caught on for day-to-day communication.

The big users were the Press, where rapid information greatly expanded the role of newspapers, and commerce, particularly passing on the prices of raw materials and stocks and shares. Governments were also very dependent on the cable system to get timely information and pass orders rapidly to other parts of their empires. Lack of such a system could have drastic consequences; in 1812, for example, a message was sent from Britain to America by sea to say that they would no longer interfere with shipping, but by the time it was received war had already broken out.[3] Fifty odd years later, when the first transatlantic cable was installed the message could be sent in a matter of minutes.

For most purposes the system was quite adequate. The dots and dashes of the Morse code could send virtually any message to anywhere in the world. However, there were some exceptions. In some far-flung places it wasn't economic to lay cables, which was expensive, particularly at sea. Only the large potential traffic had justified the enormous expense of the transatlantic cables, for example. The other important exception was ships, which once they had sailed were out of communication until they reached the next port. Could something be invented to communicate with these?

© Springer International Publishing AG 2018
J.B. Williams, *The Electric Century*, Springer Praxis Books, DOI 10.1007/978-3-319-51155-9_8

The story really begins when James Clerk Maxwell published his "Treatise on Electricity and Magnetism" in 1873 though he had been working on his ideas on and off since 1855.[4] This was one of those rare cases when theory got ahead of practice.[5]

Maxwell was a native of Edinburgh, but graduated in mathematics from Cambridge and spent most of his working life either there or at King's College in London. At the time, mathematics was considered to cover mechanics, the theory of gravitation, as well as optics, subjects which today would be described as physics. He thus had a very sound theoretical and mathematical background to begin his investigations of the work of Faraday in electricity and magnetism.

What he achieved was a set of equations based on a sound theoretical model which encompassed the laws of electric and magnetic fields propounded by Gauss, the laws of electromagnetism from Faraday, and Ampere's law. It was in this last that he made his contribution because he realized that it didn't cover the case when the magnetic field was changing. To correct this he added another term.[6] This doesn't seem much but there were far-reaching implications.

What he realized was that the equations said that it was possible to have an oscillating electromagnetic wave that propagated through free space (the air). In addition, the speed at which it traveled could be calculated from the basic electrostatic and electromagnetic constants. He measured these and arrived at a velocity which was remarkably similar to the accepted value of the speed of light. From this he concluded that light was a form of electromagnetic radiation, and that a whole spectrum should exist all the way from very low frequencies to the extremes of light.

The conclusion was that it should be possible to send a wave of energy through the air from one place to another. He had never seen these waves, and nor had anyone else, but there was a clear theoretical basis that predicted that they should exist. It was a staggering achievement, but sadly he died of cancer in 1879 at the age of 48, before conclusive proof was obtained.[7]

The practical confirmation was a surprisingly long time in coming. In the year Maxwell died a Welshman, David E. Hughes, demonstrated wireless communication, but none of his eminent audience appreciated what he had achieved.[8] Eventually the scene moved to Germany. In 1879, Heinrich Hertz turned down a suggestion of his professor, Herman von Helmholz, to investigate the predictions of Maxwell's work. However, in 1886 when he had become a professor himself at Karlsruhe he took up the challenge (Fig. 8.1).[9]

His transmitter was a Ruhmkorff coil, an induction coil similar to those now used to provide the spark for a gasoline or petrol engine, with its own make and break arrangement like a buzzer. It produced a series of high voltage signals. The output was connected across

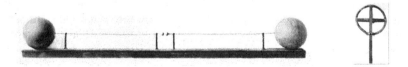

Fig. 8.1 Hertz' equipment for demonstrating wireless waves. (*Left*) Transmitter with the spark gap in the center and (*right*) a receiver with the micrometer gap at the bottom. Source: https://en.wikipedia.org/wiki/Heinrich_Hertz

a spark gap on each side of which was a large sphere. On each voltage "spike" from the coil the arc would occur and a short oscillation of power would occur between the inductance of the coil and the capacitance formed by the spheres. This would be radiated.

His receiver was simply a length of stiff wire, of the correct length for the wave he was generating, bent round in a ring. On each end were small spheres not quite touching but with a fine adjustment so that they could be brought very close together. To detect the reception he sat in a darkened room and, once his eyes had become used to the conditions, watching for the minute spark that occurred between the spheres.[10]

Amazingly, he was not only able to show that transmission took place but also that this was in the form of waves just as Maxwell had predicted. They could be reflected and refracted like light so there could be little doubt that the predictions of the equations were correct and that light was a form of electromagnetic radiation.

He had set out to show that Maxwell was right, and having proved it, thought that what he had discovered had no further interest. When his results were published in 1887, and disseminated more widely by commentators such as Oliver Lodge, the scientific community was convinced, and people began to see that there were considerable possibilities. Hertz didn't live to see them exploited as he died in 1894 of blood poisoning at only 36 years old. His contribution was to be recognized later when the unit of frequency was renamed from cycles/second to Hertz.

So many people started to experiment with the new "Hertzian" waves that there is endless scope for claim and counterclaim as to who invented what, and exactly who knew about each other's work. It is likely that the main line of development was as follows. In 1890, Eduoard Branly, who was the professor of physics at the Catholic University of Paris, published some findings on a tube full of metal filings that clumped together when subjected to the electromagnetic waves from a spark generator.[11] Importantly, the resistance of the tube dropped to a low value when this happened.

Four years later, Oliver Lodge demonstrated a system at the Royal Institution and Royal Society in London at a public lecture in honor of Hertz, who had just died.[12] He used a spark transmitter and a Branly tube as a receiver but improved it by adding a trembler which shook the contents to "unclump" them. Thus, once the transmitter signal disappeared the device, which he named a "coherer", would return to high resistance. To make the demonstration more impressive it was arranged that a bell would ring when the coherer went to low resistance. Therefore, if he pressed a key to cause the transmitter to operate, the bell would ring on the remote receiver. It was a clear demonstration of the possibilities of wireless transmission but he seemed to lose interest in the subject and didn't attempt to patent any of his ideas until 1897.[13]

One person who took a particular interest in Hertz' work was a young man of mixed Italian and British ancestry—father was an Italian administrator and landowner and his mother was Annie Jameson, a daughter of the Irish whiskey distilling family. His name was Guglielmo Marconi (Fig. 8.2).[14] In 1894, studying under Professor Righi, he became familiar with all the current work on the subject and began to try to improve the transmitters and receivers.

He set up a laboratory in the attic of his parent's country estate, Villa Griffone at Pontecchio, near Bologna in Italy, and in early 1895 was able to call his mother to a demonstration where he could make a bell ring some 9 m from his transmitter. In the spring he set up an experiment to transmit to a receiver in a barn 1.5 km distant and hidden from

Fig. 8.2 Guglielmo Marconi. Source: https://en.wikipedia.org/wiki/Guglielmo_Marconi#/media/File:Guglielmo_Marconi.jpg

view behind a hill.[15] Successful reception was indicated by firing a gun. This was the first indication that the signals didn't have to be strictly line-of-sight, and that they were reflected or refracted in some way.

Though only 20, what set Marconi apart from other experimenters was that he had a clear view of what he was trying to do: to create a commercial wireless telegraph system. As a first step, he approached the Italian Department of Posts and Telegraphs, but typically they claimed they already had a satisfactory wired telegraph network and could not see the need for a wireless one.

Undaunted, in February 1896 he and his mother, who believed in her son's work, went to England where the opportunities looked better and her family could provide him with contacts and money. The networking soon paid off as his cousin, Henry Jameson-Davis, introduced him to the consulting electrical engineer A.A. Campbell Swinton. He, in turn, wrote a letter of introduction to the Engineer in Chief of the General Post Office responsible for the telegraph system, William Preece.[16]

Surprisingly, Preece agreed to see the young man. This was Marconi's great stroke of luck; Preece himself had been experimenting, trying to produce a wireless telegraph system so that the Post Office could communicate with places not easily reached by cable telegraph such as lightships and lighthouses. Preece showed his caliber by recognizing what Marconi had achieved and encouraging him to the extent of providing his assistant, George Kemp, to help with the experiments.

First, though, Marconi needed to file a patent before disclosing the details to anyone else. Again, Jameson-Davis stepped in, introducing him to a patent agent, and on June 2, 1896 Marconi's application went in under the innocuous title of "Improvements in Transmitting Electrical Impulses and Signals, and in apparatus therefore".[17] Though this basically used the techniques already shown by Lodge and his predecessors, the real claims were in the detail improvements to make it actually function as a useful system. In particular, Marconi had realized that it worked much better if, instead of two "antennas" or "aerials" at both transmitter and receiver, one side was connected to earth and the other used as an extended antenna.

Now he could disclose everything to Preece and they set up a demonstration in July between the roofs of the Post Office building at St Martins-le-Grand and the Savings Bank department in Queen Victoria Street, some 400 m apart.[18] On September 2, 1876 the equipment was taken to Salisbury Plain and, in front of observers from the War Office and Admiralty, he only achieved a distance of a third of a mile (½ km) with a 10-foot (3 m) antenna, but extended this to 1.5 miles (2.5 km) with parabolic reflectors. Six months later, in March 1897, he returned to the same site and more than doubled the range by using kites and balloons for 120-foot (40 m) long antennas.[19]

Preece was keen to see if the system worked over water and hadn't quite given up his own ideas, so in May a trial was arranged across the Bristol Channel. While his system didn't perform well, Marconi's easily bridged the 8.7 miles (14 km), the water proving no impediment. This must have given Marconi an idea as to where the future lay. There was also more progress, as amongst the observers were some Italians who invited him to demonstrate the system in his home country.

This was too good an opportunity to miss, and the demonstration in Rome where he signaled the message "Viva l'Italia", was so successful that it led to a dinner with the King and Queen. It pointed up what he had achieved in the little more than a year since he had left the country as an unqualified school-leaver. Even better was the invitation to demonstrate the system to the Italian Navy. The resulting trial at the naval base of La Spezia successfully communicated with a warship 11 miles (18 km) away and "below the horizon".

In truth, Marconi should have been in England at this time as his company was being established as the Wireless Telegraph and Signal Company Ltd., with Henry Jameson-Davis as the managing director. The share capital came mainly from his family but, importantly, it bought his patent rights for all countries outside Italy, making him at last financially independent of his parents. He also retained 60% of the shares.

Now he could move forward with his plan. The first step, after recruiting some staff, was to set up a fixed station which was established at the Royal Needles Hotel on the Isle of Wight. With an antenna 120 ft (40 m) high, erected with the help of coastguards, he conducted trials with receivers placed in ferry boats, achieving communication at ranges of up to 18 miles (29 km). At the beginning of 1898, he went on to set up another fixed station in Bournemouth, which was lucky as the aged past prime minister, W.E. Gladstone, was staying there for a while trying to improve his health. This brought Marconi into contact with the press who were following the grand old man around like vultures.[20] (Gladstone was to die in May.)

The next stroke of luck was that the Queen was also on the Isle of Wight, but the Prince of Wales was in the royal yacht having damaged his knee. Marconi received a special commission to link them by wireless so she could receive regular news of her son's condition. In all, some 150 messages were sent between them. The Queen was reported to be delighted.

With the Italian navy adopting his system Marconi knew he was on his way, but still most of his time was taken in demonstrations. The one for the House of Commons only carried across the river, but it was good publicity. Lord Kelvin, the famous physicist, visited the Needles station and, though skeptical to begin with, ended up insisting on sending a telegram by wireless and paying for it. Though this was the first wireless telegram it conflicted with the Post Office's monopoly and so was probably illegal.

Next came a trial for Trinity House, who were responsible for the lighthouses and lightships. The test was conducted between the South Foreland lighthouse, near Dover, and the East Goodwin lightship, a distance of some 12 miles (20 km). The system was left in place and proved its worth in early 1899 when heavy seas damaged the lightship and it could report this by wireless. Two months later, the German ship "Elbe" ran aground on the Goodwin Sands in dense fog. The Ramsgate lifeboat was summoned with a wireless message. A month later it was the lightship itself that again was in difficulties when it was rammed, and was able to send out the very first distress call.

Marconi was thinking ahead and deliberately made the South Foreland transmitter more powerful than was necessary. As soon as he obtained permission, he set up another station just across the Channel in France at Wimereaux near Boulogne. This attracted the attention of the French navy, and not to be outdone, the British and American navies. The Royal Navy conducted successful trials at ranges up to 45 miles (75 km), but was reluctant to actually part with any money to buy a system.

At the turn of the century, Marconi had made tremendous progress in publicizing his equipment and convincing the world that it did work, but this hadn't brought much return and his company was close to financial collapse. This didn't deter him at all. With his high profile it made sense to change the name to Marconi's Wireless Telegraph Company. Also the large number of enquiries for marine systems defined the future direction and led to the setting up of a subsidiary to deal with them named Marconi International Marine Communication Company. This established offices around the world.

On the technical side, Marconi also had a trick up his sleeve. In April 1900, he applied for what would become his famous 7777 patent.[21] The problem he had been having was that, when a transmitter was operating, the signal covered such a wide band of frequencies that it blotted out anything else in the area. What he had now started using, and wished to protect, was a system of loosely coupling the spark system to the antenna and tuning that as well. This was repeated at the receiving end. This largely cured the problem and, in addition, because the transmitted signal didn't die away as quickly and was more concentrated at a single frequency, a more powerful wave was radiated. Of course, this traveled further and extended the effective range of the stations.

The company began to make commercial progress with an order from the Royal Navy and installations in merchant ships. These included the German "Kaiser Wilhelm der Grosse", built to capture the Blue Riband for the fastest Atlantic crossing from the Cunard line's "Lucania", which was also equipped. There were clear signs that the system was beginning to be adopted.

It was then that Marconi's nerve showed itself. He stunned some of his fellow directors with his plan to bridge the Atlantic by wireless. This was a tremendous gamble, both financially and technically. He saw that wireless communication would only be truly successful if it spanned the world, but the cost of building a high-power station would be considerable.

Light travels in straight lines, and so basically do radio waves. It was thus thought that the curvature of the Earth would make long-distance communication impossible. Marconi was convinced this was wrong, and he had good reasons to be so. In addition to those early tests which had suggested that some form of reflection or refraction took place, the signal from the French station at Wimereaux had been picked up in the factory at Chelmsford some 80 miles (130 km) away, well beyond a 'line of sight' range.

As a preliminary, Marconi set up a station on the Lizard peninsular in Cornwall and successfully established two-way communications with a station on the Isle of Wight 186 miles (300 km) away. His antenna would have needed to be more than a mile (1.6 km) high if the signal went in a straight line, but it was nothing like this. It was further proof that the signal was bending around the Earth.

Work began on a massively powerful transmitter at Poldhu, a headland a little nearer Land's End, and almost as far to the west of England as possible. The antenna was an inverted cone of wires 400 ft (120 m) in diameter and 200 ft (60 m) high. It worried his site engineer, R.N. Vyvyan, who thought it too fragile for the exposed position. Nevertheless, Marconi went ahead and took Vyvyan to America to produce a duplicate at Cape Cod. This was all conducted with as much secrecy as possible as he didn't want anyone to know what they were attempting.

They were encouraged when a test with the partially completed antenna at Poldhu was picked up by another station in Ireland 225 miles (360 km) away. However, Cape Cod was more than ten times distant. As a fallback position he started to consider the nearest point in north America which was St John's in Newfoundland, still a British colony, at a range of only 1800 miles (2880 km).

It was just as well, because Vyvyan had been right and the Poldhu antenna blew down, though Marconi immediately put in hand a temporary replacement while a more permanent stronger structure was built (Fig. 8.3). Winter was fast approaching and he felt that he

Fig. 8.3 The Poldhu station, with the temporary antenna used for the first transatlantic wireless test. It redeployed two of the masts from the original circular one that blew down in a storm. Source: http://www.marconicalling.co.uk/introsting.htm

should seize the moment. He set off for Liverpool to catch a ship only to receive a telegram to say that the Cape Cod antenna had also blown down narrowly missing killing Vyvyan.

Marconi was already on Plan B as the ship he and his team took was heading for Newfoundland. There they put out a cover story about ship-to-shore wireless to explain the kites and army balloons that they were using. On December 9, 1901 they connected up all the equipment and cabled Poldhu to transmit the Morse letter "S", dot-dot-dot, for 3 h every day. Transmissions began 2 days later but at St. John's a balloon was lost in bad weather, and they were unsure if they had heard anything.

The next day the wind was even stronger and a kite was lost. The gale was swinging the antenna violently, but they persevered. Then at 12.30 pm Marconi heard, through the atmospherics, the faint dot-dot-dot from Poldhu. He passed the earpiece to his assistant, Kemp, who listened. He could hear it too. Twice more that day they heard the signal, but then the weather worsened and they had to stop.

Marconi now had to decide whether to announce this achievement to the world, which might bring ridicule, or say nothing and miss the chance. After cabling the news to London he waited until December 16 before he announced it to the press. The reaction was mixed. While he was feted in America, some English scientists remained skeptical and tried to find other explanations. The Canadian government wanted him to establish a permanent station. The local cable company wanted to sue him for breaking their monopoly.

Marconi didn't stand still, and before returning to England he signed an agreement with the Canadian government to set up a station at Cape Breton in Nova Scotia. He had barely time to brief his fellow directors and attend the Annual General meeting of the company before taking the American Lines "Philadelphia" back to America. With him was a team of wireless operators and engineers who fitted a large antenna system to the masts of the ship. Inside, the reception of the messages was to be recorded on a Morse inker tape and witnessed by the Captain.

In daylight, they received signals from Poldhu up to a range of 700 miles (1120 km); at night, he recorded messages at 1550 miles (2500 km) and, in conditions that left no room for doubt, picked up the three dots of the Morse code 'S' at 2100 miles (3380 km). There could now be no doubt that long-range wireless telegraphy was possible.

These achievements had all been made using coherer detectors, but they were not always satisfactory, particularly at sea. Always staying a step ahead, Marconi came up with a magnetic detector and applied for a patent. The detector used a band of iron wires, kept moving by a clockwork mechanism, which passed under a magnet and then through a pair of concentric coils. One of these was coupled to the antenna and the other to an earphone. When a signal was received the magnetic field in the wire changed and a click heard. It was a rugged and reliable system.

Steadily the company's systems were being fitted to ships around the world. A factor in the Japanese Navy's victory over the Russians in 1905 was that its ships were equipped with wireless and the Russians were not. Despite the progress, such as the commencement of the transatlantic service, the enormous cost of setting up all the stations was a drain on the company. By 1907, the company was again in financial difficulties and shed large numbers of people, including the managing director. Marconi, though no administrator, had to take over to save the company. It survived and things began to look up as the returns from all the installed systems began to flow.

In 1909, came a huge accolade: Marconi won the Nobel Prize for Physics, "in recognition of contributions to the development of wireless telegraphy". Surprisingly, it was shared with Ferdinand Braun, a professor at Strasbourg University, then in Germany. As there were a number of others who had equally made a contribution, perhaps this was for political reasons. It was embarrassing for both men as the parallel work Braun had been carrying out had led to the setting up of the rival German company Telefunken. There had been considerable commercial competition between the two which had become a proxy for the growing strains between Britain and Germany. Despite many attempts, and Marconi's acceptance that he had "borrowed" ideas from Braun, the Germans were unable to break the Marconi patents and the company's near monopoly on marine systems.

In 1912, came the "Titanic" disaster. One of the two Marconi operators, Jack Phillips, stayed at his post transmitting the distress calls, and went down with the ship. It was only the reception of these by the "Carpathia" that led to the saving of many lives. The "Californian" had been even nearer, but the single wireless operator had been on duty all day and had gone to bed, and the SOS was not heard. By a further turn of fate, Marconi and his wife had been invited to sail on the "Titanic's" maiden voyage, but business commitments for him, and the illness of their son for his wife, meant that neither were on board.

One of the results of the disaster was that ships had to keep a 24 h wireless watch for distress signals. It went without saying that a wireless system was now almost essential. The number of installed systems rose sharply from 1912 onwards. This was good business for the Marconi company which dominated the field (Fig. 8.4).

Marconi is one of the rare examples where the name most associated with a development is undoubtedly that of the person who really got it going. His critics claimed that he never invented anything, and there is little doubt that most of his equipment was derived from other people's work. But that misses the point. He refined their ideas and built them

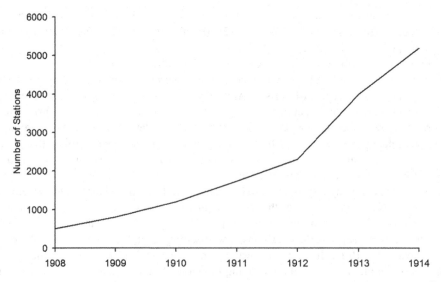

Fig. 8.4 Rise in the number of fixed- and ship-based wireless stations, 1908–1914. Source Author.[22]

into a working system, publicized it and persuaded the world that it was needed and to accept it; all that in less than 20 years. That was his achievement. It was his clear-sighted vision that led the company to take the huge technical gambles and then, when they could have rested on their laurels, to seek to bridge the Atlantic. Through all the difficult times he never wavered.

Oliver Lodge disputed Marconi's 7777 patent, claiming that he had invented the tuned coupling method, but while it is hinted at in his patent, it is not clear-cut like Marconi's.[23] The patent examiners had allowed Marconi's patent, not recognizing that there was a conflict. Wisely, rather than indulge in litigation, the company bought it from Lodge, avoiding any problems. Lodge always tried to denigrate Marconi, but there is little doubt that he would never have shown the same drive to overcome the obstacles, and he really didn't understand the other man's skills.

From the original skepticism of the 1890s, Marconi had not only proved that wireless telegraphy was feasible, but had built a business and the foundations of a whole industry. The idea that communication could be made through the ether was now accepted, and would never go away. That was to have profound consequences. Though at the time wireless telegraphy was still the preserve of governments and business, or a plaything of the rich on liners, the possibilities were there for further development.

Soon the war, as so often, would drive those changes, and give rise to electronics. This would spawn radio broadcasting, television, radar and much else. It would lead on to computers, mobile cell phones, the internet and all those things we take for granted. However, electronics, that child of electricity, is a complete subject in itself and that has to be handled elsewhere.

Marconi died on July 20, 1937, aged 63, and he was buried at Villa Griffone. In a fitting tribute, wireless stations throughout the world stopped for 2 min and the ether was, temporarily, as silent as it had been before he began.

NOTES

1. de Cogan D. & Baldwin T. Technological innovation and financial stagnation (the growth of international telegraphy 1866–1900: A British perspective), available at: http://www.dandadec.co.uk/page32/page41/page41.html.
2. Calculated from de Cogan & Baldwin.
3. Wheen A. *Dot-Dash to Dot.Com: How Modern Telecommunications Evolved from the Telegraph to the Internet*, p. 1.
4. DNB, James Clerk Maxwell (1831–1879).
5. The other major example is Einstein's relativity. Albert Einstein acknowledged his debt to Maxwell and used his work as a starting point.
6. Hansen M. Maxwell's equations, available at: www.artofproblemsolving.com/.../Maxwell%27s%20Equations.pdf.
7. DNB, James Clerk Maxwell (1831–1879).
8. Information about biography: Hughes I. and Evans D.E. Before we went wireless, available at: http://www.imagesfromthepast.com/BWWW.html; Royal Society, Obituary of Oliver Lodge, available at: http://rsbm.royalsocietypublishing.org/content/obits/3/10/550.
9. NNDB, Heinrich Hertz, available at: http://www.nndb.com/people/419/000072203/.

10. New World Encyclopaedia, Heinrich Hertz, available at: http://www.newworldencyclopedia. org/entry/Heinrich_Hertz.
11. Science Museum Online Science, Branly, Édouard Eugène Désiré, available at: http://www. sciencemuseum.org.uk/online_science/explore_our_collections/people/branly_edouard.
12. DNB Lodge, Sir Oliver Joseph (1851–1940).
13. British patent GB189711575.
14. DNB, Guglielmo Marconi (1874–1937).
15. DNB gives the distance as 1.5 km but the Biography for the Nobel Prize gives 1.5 miles. It would seem more likely to be in kilometers in Italy and the DNB uses Italian sources. However, other sources say 2 km.
16. Marconi Collection Archive, Marconi meets Preece, and succeeding pages, available by selecting date at: http://www.marconicalling.co.uk/introsting.htm.
17. British patent GB189612039.
18. Marconi Collection Archive. Other sources have 300 yds, but as the Marconi Archive have the records they are likely to be the most reliable.
19. The rest of this story largely comes from the Marconi Collection Archive.
20. E.g. *The Times*, April 20, 1898.
21. Patent GB190007777.
22. Data from *Wireless World* April 1961, Golden Jubilee Review of 1912.
23. Patent GB189711575.

9

Healthy? Early Medical Electricity

I need hardly recall to mind, that until quite recently, to venture to speak of electricity as a curative power was pretty certain to result in the speaker being branded as little better than a quack.

Herbert Tibbets, MD, A Handbook of Medical and Surgical Electricity, 1877

On January 17, 1803 George Foster was executed for the murder of his wife and child who he had drowned in the Paddington Canal in London.[1] If it hadn't been for what happened next, he wouldn't have rated a footnote in the history books. After hanging for the usual time, he was cut down and taken to a nearby house where he was subjected to the "galvanic process" by Professor Aldini. The ends of a large battery of 120 pairs of plates of zinc and copper were connected to two electrodes. These were inserted into Foster's mouth and ear, giving dramatic results. Aldini noted that "the jaw began to quiver, the adjoining muscles were horribly contorted, and the left eye actually opened". When one rod touched the rectum, the whole body convulsed; indeed, the movements were "so much increased as almost to give an appearance of reanimation".

This whole gruesome process was supervised by Dr. Thomas Keate, then president of the Royal College of Surgeons in London. It was supposed to be a serious scientific experiment to investigate whether the procedure might be useful to revive people who had drowned. The report in *The Times* made a strong impression on the readers and is thought to be the genesis of Mary Shelley's *Frankenstein*. The dividing line between serious science, quackery and fantasy was very slim.

Giovanni Aldini was the nephew of Professor Luigi Galvani who, in 1791 in Bologna, had published the results of his experiments which showed that an electric impulse could cause a frog's leg to twitch. He ascribed the cause to intrinsic "animal electricity" which was generated by the brain and ran along the nerves. Down the road in Pavia another professor, Alessandro Volta, had received the results eagerly but began to doubt Galvani's interpretation. He thought that, instead of demonstrating the existence of intrinsic animal electricity, these muscle contractions were the result of electricity generated by two dissimilar metals.

© Springer International Publishing AG 2018
J.B. Williams, *The Electric Century*, Springer Praxis Books, DOI 10.1007/978-3-319-51155-9_9

In a sense, they both turned out to be partially right. What had been demonstrated was the connection between electricity and functions of the body. Something else came out of this as Volta then made a "pile" of pairs of discs of copper and zinc with salt solution between them and showed that they did generate a continuous form of electricity. The battery, which was to have a great impact on the development of electricity, was born. It is still called a "pile" in many languages.

Ever since the discovery of electrical effects, people have been trying to use them for influencing, and hopefully curing, ailments of the body. The shocks from torpedo fish and the gymnoti or electric eel had been used for hundreds of years to numb pain. Storm van 's Gravesande, who was the Governor of Surinam in 1754, wrote that: "people who to some degree had gouty pains, had been completely cured two or three minutes after contact with the torpedo".[2] In Berbice and Demerara the colonists used to keep live gymnoti in a tank for use by their plantation workers, who had great faith in the power of the fish's shock to cure rheumatic and paralytic afflictions. They weren't the only ones who believed in it. As late as 1850, European doctors in Guiana were using the shocks to treat rheumatism.

After the invention of generators that produced static electricity and the Leiden jar to store it in the second half of the eighteenth century, the path was open for the "scientific study" of the effects of electricity on the human body.[3] It wasn't long before it found its way into hospitals. By 1780, John Birch, a surgeon at St. Thomas' Hospital in London was conducting what he called "experiments in medical electricity".[4] He explains his approach: "The applications of the electric fluid to the diseases of the human body may be comprised under three heads: the first, under the form of radii, when projected from a point; the second, under the form of a spark, when many of these radii are concentered on a ball; and third, under that of a globe, when many of these sparks are condensed in a Leiden jar. Now, to each of these heads a specific action belongs: the first, or fluid state, acts as a sedative; the second, concentrated state, acts as a stimulant; and the last, condensed state, as a deobstruent."[5]

To the modern mind it makes no sense at all, but he gives many examples of "cures" effected by his methods. While some were probably the result of the placebo effect, and other patients would have got better anyway, it isn't possible to dismiss them all. The examples of dealing with paralysis of a limb (a frozen limb rather than one caused by a break in the spinal column) seem quite plausible. On the other hand, putting electric discharges through men's testicles and their heads strike one as downright dangerous. One can only assume that the energy that the equipment produced wasn't sufficient to do any real harm (Fig. 9.1).

These electrostatic discharges to or from the body—known as Franklinism, after the work of Benjamin Franklin—were often in the form of the "air bath". The patient sat on a stool which was insulated from the ground. One end of the static electricity generator was connected to their clothing or held in the hand while the other was connected to ground. The operator then used a probe connected to ground to draw sparks from the appropriate part of the body. Sparks were drawn from many different parts of the patient for amazing reasons including barrenness in women (infertility) and impotence in men.

With the arrival of batteries, a whole new range of possibilities opened up. As Aldini had shown, electricity could cause a muscle to twitch but only when it was applied or removed. If it was left there, nothing happened. While it was used as an overall "bath" or as a local stimulus it is unlikely that these did much except give the patient a good feeling. There were specific areas where it was much more useful.

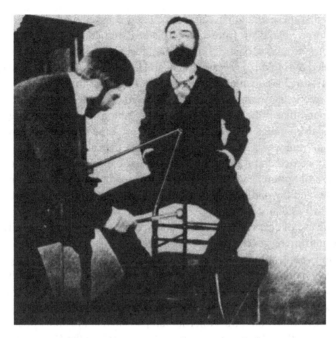

Fig. 9.1. Drawing sparks from the genitalia. Source: L.A. Geddes, Electrical treatments, *The Physiologist*, 27:1, Suppl., 1984.

Either applied directly or in the form of a heated wire electricity could be used to destroy tissue such as skin tumors, or to form clots to prevent bleeding. This was very useful during surgery. A curious use was to assist the joining of broken bones that didn't want to knit together. These uses have lived on and are still used to some extent.

In the nineteenth century, there was a particular problem with the use of chloroform as an anesthetic in that it could stop the patient's breathing or heart. It was found that a carefully administered shock could twitch the respiratory muscles and get the breathing and circulation going again. It was recommended that the surgeon should have a battery handy in the operating theater to deal with this eventuality.[6]

In the latter half of the nineteenth century with the introduction of induction coils another form of electrical treatment became available, which was known as Faradic. The name derived from Michael Faraday because the method involved an electromagnetic component. Where a Galvanic current produced a response in a muscle or nerve only when it was switched on or off, the stream of "spikes" from the induction coil (without a spark gap) produced tetany or the constant contraction of the muscle.

Once again very suspect experiments on whole body and local Faradism were carried out in an attempt to treat such things as dyspepsia, hysteria, neuralgia, rheumatism, paralysis, temporary constipation, and insomnia.[7] Nettie Rosch of White Plains, New York, was a "hopeless paralytic", having lost the use of her lower limbs.[8] She was claimed to be cured by Dr. Thomas W. Topham who gave her two months of "electrical treatment". She was reported to be able to walk home from the station.

It is doubtful that these would pass modern evidence-based medicine tests. More worryingly, trying to use Faradism to restart the heart could cause it to go into fibrillation where the muscles around it contract in an irregular and unsynchronized fashion. How many patients died during this process is unclear. However, in 1889 J. A. McWilliam discovered that bursts of Faradism at something like 70 to 120 to the minute could get the heart beating correctly. However, this wasn't followed up at the time.

It was also found that if the frequency of the excitation was increased, either from an induction coil or from a generator (some Faradism used the mains if they were supplied with alternating current which sounds very dangerous), then above about 4 kHz the muscle contraction effect ceased and it merely heated the muscle or tissue. This was found to be useful in the case of spasms and a number of other conditions. When the frequency was raised to 1 MHz, and beyond, the devices effectively became radio transmitters and the process was known as diathermy.[9]

Of course, the use of electric devices to influence aspects of the body spilled out of the medical world into more general use in the latter half of the nineteenth century. These varied from things that might possibly have been of some use to downright quackery. Electro-magnetic hairbrushes were supposed to stimulate hair growth, while the Electropathic Belt (for suffering men and women) was for neuralgia, rheumatism, indigestion, sleeplessness, ladies' ailments, etc.[10] It was claimed to impart new life and vigor.[11]

Popular were infra-red and ultra-violet sunlamps which could apply heat and probably did give some relief, then there were fancy machines such as the Rotosurge which came with a whole range of glass electrodes for treating different parts of the body. These would glow blue, violet or pink when energized with high voltage at high frequencies and would emit streams of sparks when in contact with different parts of the body.[12] It is doubtful that in unskilled hands they effected many cures, or would pass any sort of investigation today.

It didn't take long for the idea to take hold that a large bolt of electricity across the chest could kill, and in America this led to the development of the Electric Chair. However, General Electric wanted to understand why it was lethal and what could be done about it. Their concern was to protect their business selling electrical equipment.

They funded two electrical engineering professors, William Kouwenhoven and Guy Knickerbocker from Johns Hopkins University in Baltimore, who tested the phenomenon by shocking stray dogs to death. During these experiments, they noticed that a second AC shock could sometimes bring an electrocuted dog back to life.[13] This work inspired a pioneering cardiac surgeon, Claude Beck, at the University Hospitals of Cleveland. He tried using AC directly onto the exposed hearts of animals he had put into ventricular fibrillation.

Then, one day in 1947 a 14-year-old patient's heart stopped during surgery. Beck had his research apparatus brought up from the hospital's basement. This simple defibrillator consisted of a transformer to isolate the patient from the 110-volt AC mains supply, a variable current-limiting resistor, and two metal tablespoons with wooden handles so that he could deliver the pulse to the exposed heart. Though the first shock failed, the second brought the lad back to life, and the story was soon all over the newspapers. Although this did cause damage sometimes, the method was used for some years, though no-one really knew why it worked.

In Europe, a different method was in use, using the discharge of a capacitor, which harped back to the early work. Gradually refinements were made and it was found that a decayed sine wave worked better. The idea behind this was that a reversed current at a lower level restored the cells that had been hit by the first pulse of forward current.

General development of electronics, and the lower power needed by this process, meant that it was possible to build portable units and fit them in ambulances, which led to the saving of many lives. However, they had to be operated by skilled personnel as it was necessary to check what type of heart problem the patient was suffering from before administering the shock. The final form was a device that was small and portable and able to make the decisions for the user, allowing its deployment in unskilled hands by giving them spoken commands. Thus the elements of electricity, electronics and medical practice have come together to produce a life-saving device.

Of course, it was only a step from there to have a device that kept the heart beating in a regular rhythm when it failed to do so by itself. This was the heart pacemaker. A box of electronics became small enough to be inserted inside the body and pulse the heart regularly to maintain its beat. It was a masterpiece of electronics and batteries and finding materials that were compatible with the body. Many people owe their ability to have a normal life to that magic box.

Other electrical treatments within medicine became part of physiotherapy. Faradic currents were used to re-educate muscles that had atrophied or been injured. The use of high frequencies became Shortwave Diathermy and was used to deliver heat to internal tissues to promote healing. These were very popular in the middle part of the twentieth century but their use, though still present, has declined.

A variation of the original electrical methods is TENS (Transcutaneous Electrical Nerve Stimulation). This uses electrical pulses, which can be focused on a particular area, to give short-term relief of acute and chronic pain by blocking the pain signals from reaching the brain. It is believed to work by increasing the quantity of endorphins that are released, and functions without resorting to medication. Technology has reduced the necessary equipment to a small, handheld portable device. Because the mechanism is not understood, and it doesn't seem to work in all cases, the high hopes for its use haven't been realized.

*

On November 8, 1895 Wilhelm Roentgen, a professor at the University of Würzburg, was having a quiet Friday afternoon in his laboratory when he could get some experiments done without interruptions. He was exploring the fluorescence of barium platinocyanide, which was pasted onto a piece of card, when subjected to the rays coming from a Crookes tube, the early forerunner of the cathode ray tube as used in older television sets. This in turn was energized with a Ruhmkorff coil, an induction coil similar to those now used to provide the spark for a gasoline or petrol engine, with its own make-and-break arrangement like a buzzer. Because the glass of the tube was itself showing fluorescence, Roentgen wrapped it in black cardboard so that he could see the weak fluorescence of his target.[14]

What surprised him was that, despite checking that the shielding was efficient, the barium platinocyanide was faintly glimmering though more than a meter away. He checked and rechecked his result to be sure that this was not due to any stray light, or to the cathode rays from the tube which were known not to travel that far through air.

He kept his discovery secret until he was quite sure and had investigated the phenomenon thoroughly. He didn't want to make a fool of himself by announcing something that would turn out to be wrong. As an experienced physicist, over the next seven weeks he conducted a carefully planned set of seventeen experiments to establish the properties of what he had discovered. So that he could work without interruption, it is said that he ate his meals in the laboratory and even had his bed moved there.

He was able to show that whatever was exciting the barium platinocyanide would pass through many materials, but was impeded by dense substances and thick metals, particularly lead. To provide evidence of the effects of his rays, he substituted a photographic plate for the barium platinocyanide panel. On December 22, he asked his wife to come to the laboratory and persuaded her to let him pass his rays through her hand for 15 min and catch the result on a photographic plate. It was the first medical X-ray, so named by him as he was unable to determine exactly what it was.

This was the final piece he needed to write a short paper, "On a new kind of rays". He handed this to the secretary of the Würzburg Physical Medical Society on December 28 for immediate publication and arranged a presentation for January 23, 1896. By New Year's Day he had some prints of the paper and sent them to fifty colleagues in Germany and Austria.[15] The recipients immediately recognized the enormous importance. The dissemination of the news was amazingly rapid in an era before modern instant communication.

From one of the copies he had sent out a report found its way into the Vienna paper *Die Presse* on Sunday January 5. Unfortunately, they misspelt his name as Routgen and that propagated through the papers which picked up the story. One of these was the *London Evening Standard* which carried a report on the evening of January 6. The next day, other papers had the story that the "new light" of this sensational discovery showed metal weights inside a wooden box and the bones inside human flesh (Fig. 9.2).[16]

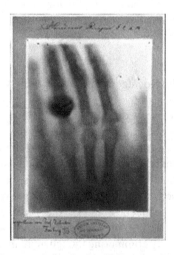

Fig. 9.2. X-ray of Bertha Roentgen's hand, clearly showing her rings. Source: http://www.medcyclopaedia.com/library/radiology/chapter01.aspx

Due to the time difference, the *Sun* in America managed to print a report also on January 6, and it went right across the States on the following day. Ironically, it only appeared in the local Würzburg paper on January 9. On January 13, Roentgen presented his work to Emperor Wilhelm II in Berlin and the wisdom of his not announcing anything until he was thoroughly prepared showed itself. Finally, on January 23, 1896, he gave his oral presentation to the Würzburg Physical Medical Society, who must have been feeling a bit left out.

Within the next few weeks, due to their longer publication cycles, the scientific journals began to catch up, all the main ones publishing some sort of report. In America, Nicola Tesla, the electrical inventor, investigated how the characteristics of the X-rays changed with variations of the gas tubes and electrical generators, and published his results in March and April.

The publication of the photograph of Roentgen's wife's hand set many in the medical world thinking, and they weren't slow to act. In America, Dr. Gilman Frost was lucky to have a brother who was a professor of physics. On February 3, he brought a patient, Eddie McCarthy, to the laboratory and they took an X-ray of his damaged wrist which clearly showed the broken ulna.[17] In March in Birmingham, England, Dr. Hall-Edwards passed X-rays right through the human body and obtained a photograph of the backbone showing the vertebrae and spinal cord in full detail.[18]

Within 6 months of the original announcement of the discovery, X-ray equipment was installed in several London hospitals.[19] The speed was partly due to the presence already of electrical departments which almost certainly possessed the necessary induction coils. They could easily obtain Crookes tubes from scientific apparatus suppliers, and photographic plates were commonly available. Thus it was simple to amass the necessary equipment.

Just because some were quick to take up the possibilities didn't mean that the conservative medical profession as a whole accepted X-rays with open arms. Some did not want to accept it at all, regarding it as quackery, and even when they did it was grudgingly. Those who ran the hospitals were far from convinced, and it was normal to find the X-ray department in some tiny cellar, accessed by the darkest, steepest stair.[20] Often these places were so damp that the first task of the day was drying the apparatus.[21]

In the next few years, as it moved from the experimental to the commonplace X-rays gradually gained acceptance. The British Army was quick to take up X-rays and in 1898 was using them under battle conditions in a mud hut on the banks of the Nile, using volunteers pedaling a bicycle to charge the batteries. In 1903, the King and Queen opened a new outpatient department at the London Hospital which was fully equipped with X-ray apparatus in its electrical division.

Gradually, the Crookes tubes were modified to be better and more reliable generators of X-rays, but it wasn't until 1913 that a really satisfactory device was available. This was produced by William Coolidge in America, and was subsequently named after him. The essential difference was that it used a heated cathode like a vacuum tube or valve and the electron stream was directed onto an anode specifically designed to produce the X-rays. One of its great advantages was that the intensity and energy of the X-rays could be controlled independently.[22]

However, at first, the dangers of X-rays were not appreciated. In March 1896, A.A. Campbell Swinton—the same man who had helped Marconi along his way—set up a laboratory to take X-ray photographs. One of his first clients was a man with a bullet in his head. The bullet was successfully located but the patient's hair began to fall out and he threatened to sue Swinton. However, the hair soon grew again and nothing came of the threat. The "Electrical World" of June 1896 even suggested that X-rays might be used as a substitute for shaving.[23]

In the various X-ray departments the operators began to get problems with their hands and soon cancers appeared; some foolishly just carried on with the result that they died. It was soon apparent that this stuff was dangerous, and that the doses should be strictly controlled not only for the patient, but even more importantly for the operator, who was repeatedly exposed and must take precautions to screen themselves from the rays. The only effective protection is distance and the metal lead. Even today, operators depart to a safe distance and behind a screen to take the actual X-ray.

The up-side was that the X-rays could be used to attack cancers as well. Early in the twentieth century it began to be used to try to cure skin cancers and some successes were obtained. The difficulties were that the early tubes were very variable in their output and so it was very difficult to know exactly what dose was being used. Many subjects were injured by operators who did not really know what they were doing and, frankly, used their patients as guinea pigs. It still hadn't been resolved who was responsible, the doctor or the operator. On one side there was the doctor who knew nothing about X-rays and on the other the X-ray physicist who knew nothing about medicine.[24] Then, of course, the subject attracted the quacks.

By 1906, X-ray therapy was beginning to advance beyond the experimental stage and the Roentgen Society published its guidelines: "The Uses of the Roentgen Rays In General Practice" which led to a more general employment of X-rays within the medical profession and the control of who could use them. At the same time, there was greater understanding of the dangers and limitations, as well as the advantages.

From these shaky beginnings, X-rays grew to become the great tool for examining the body internally and were extensively used for examining broken bones and screening lungs for tuberculosis. It was also to have a new lease of life in CT (computed tomography) scanners. The medical profession would be very different today without their help. At the same time, more carefully judged use of them for therapeutic purposes led to a useful weapon in the doctor's armory when it came to dealing with many sorts of cancers, particularly to back up other forms of treatment or for use when all else had failed.

A curious aside to the use of X-rays was their scientific value in examining the structure of crystals. This led in 1951 to the investigation into the form of the DNA molecule. The unraveling of the double helix structure over the next few years was based on X-ray crystallography, and in particular the work of Rosalind Franklin at King's College in London. Unfortunately, she died before her name could be added to those of Crick, Watson and Wilkins for the Nobel Prize for Medicine in 1962.

Surprisingly, after the early high hopes, by the end of the twentieth century electrical methods were not greatly employed for medical therapy, except perhaps for defibrillators and pain relief. The main use was for investigation and diagnosis. Perhaps this is unsurprising in a medical profession dominated by surgery and pharmaceuticals, but it seems that the possibilities offered by the electrical properties of the body have never been fully exploited. The areas where electricity, as well as its offshoots electronics and computing, was able to make a real contribution was in diagnostics.

NOTES

1. Parent A. Giovanni Aldini: From animal electricity to human brain stimulation. *Canadian Journal of Neurological Sciences*, 31: 4, November 2004, also available at: http://www.bius-ante.parisdescartes.fr/chn/docpdf/parent_aldini.pdf; He states that this took place on January 18, 1803, quoting the Newgate calendar, but he refers to *The Times* of January 22, 1803 which clearly states that this took place on the previous Monday, which was also stated as the day when his sentence was reported in *The Times* on the previous Saturday. A look at a calendar for that year shows that Monday was the 17th. Aldini wrote that it was the 17th. It would seem more likely that the Newgate calendar is muddled rather than everybody else.

2. Gadsby G. Electroanalgesia: Historical and Contemporary Developments. PhD Thesis, De Montford University, 1998.

3. These machines were cruder predecessors of the Wimshurst machines common in school physics laboratories. They all worked by rubbing suitable surfaces with a "brush" to generate static electricity and were thus known as friction machines; The Leiden (or Leyden) Jar, an early form of capacitor, was invented by several people in the mid-18th century. The name comes from the town in Holland where Pieter van Musschenbroek, one of the inventors, worked. Presumably no-one could pronounce his name!

4. Birch J. A letter on the subject of medical electricity. In G. Adams, *An Essay on Electricity*, 5th edition, p. 507.

5. Birch, p. 509.

6. Geddes L.A. Monograph on early electrical treatments. *Supplement to The Physiologist*, 27:1, 1984.

7. Geddes, Chapter 4.

8. *New York Times*, December 25, 1900.

9. Geddes, Chapter 5.

10. Jenkins J.D. Quack medial apparatus, available at: http://www.sparkmuseum.com/QUACK.HTM.

11. Gordon B. *Early Electrical Appliances*, p. 28.

12. Gordon, p. 27.

13. Kroll M.W., Kroll K. and Gilman B. Idiot proofing the defibrillator. *IEEE Spectrum*, November 2008.

14. British Library, Roentgen's discovery of the X-ray, available at: http://www.bl.uk/learning/cult/bodies/xray/roentgen.html.

15. Linton O.W. News of X-ray reaches America days after announcement of Roentgen's discovery, *American Journal of Roentgenology*, 165, 471–472, 1995.

16. *Guardian*, January 7, 1896.

17. Spiegel P.K. The first clinical x-ray made in America – 100 years. *American Journal of Roentgenology*, 164, 241–243,1995.

18. *Guardian*, March 15, 1896.

19. Goodman P.C, The x-ray enters the hospital. *American Journal of Roentgenology*, 165, 1046–1050, 1995.

20. Goodman.

21. Thomas A.M.K. The development of radiology from the discovery of x-rays in 1895. *The Journal of The British Society for the History of Radiology*, No 23, November 2005.

22. Frame P. Coolidge X-ray tubes, available at http://www.orau.org/PTP/collection/xray-tubescoolidge/coolidgeinformation.htm.

23. Oldnall N. A brief history of x-rays, based on B. Bowers, X-rays: Their discovery and application, available at: http://www.e-radiography.net/history/general.htm.

24. Timothy N.D. X-ray therapy and the early years, 1902–1907. *The Journal of The British Society for the History of Radiology*, No 28, November 2008.

10

Portable Power: Batteries

Police arrested two kids yesterday, one was drinking battery acid, the other was eating fireworks. They charged one and let the other one off.

Tommy Cooper, British comedian

John Frederic Daniell was so outstanding that he was elected a member of the Royal Society at only 24 years of age.[1] He was born in 1790 and well educated by private tutors, and showed a distinct interest in science at an early age.[2] Before he was out of his teens he had worked in a sugar refinery belonging to a relative of his mother's improving its processes. In the next few years he became particularly interested in geology, assembling a fine collection of rocks. Through Professor Brande he became involved with the Royal Institution.

Not content with that he started work on meteorology, leading in 1820 to his development of the dew-point hygrometer, which became the standard method of measuring humidity. This was followed by work on the atmosphere of hothouses and the importance of controlling humidity as well as temperature.[3]

In 1825, he became a director and the superintendent of the grandly named Imperial Continental Gas Association and traveled all over Europe to interest towns and districts in gas lighting.[4] It was quite successful with installations in Berlin and other cities and continued to exist until 1987 when it split into two, one part being Calor Gas.

Despite his lack of academic background, in 1831 he switched to academia, becoming the professor of chemistry at King's College in London, and a respected lecturer. He was a friend of Michael Faraday and through him became interested in electricity, so he soon put his newly-equipped laboratory at the college to use. In 1836, he produced the Daniell cell which was the first reliable and practicable battery.

Of course, Alessandro Volta had invented the basic device in 1800, but though it was some use for scientific work, it suffered from a fundamental drawback in that it rapidly lost power. This was due to a thin film of hydrogen bubbles forming over the positive electrode, in a process known as polarization. The effect is to increase the internal resistance and so reduce the effective voltage of the cell. Rapidly it becomes unusable.

© Springer International Publishing AG 2018
J.B. Williams, *The Electric Century*, Springer Praxis Books, DOI 10.1007/978-3-319-51155-9_10

Daniell's cell solved this problem and could give a continuous output of 1.1 volts for an extended period. It didn't produce hydrogen and so didn't suffer from polarization. This was achieved by having a cylindrical copper vessel that served as one electrode. Inside this was a porous earthenware container or partition with a zinc rod which was the other electrode. In the space between the copper and the porous pot was a solution of copper sulfate, which was kept saturated by extra crystals. The porous pot, which contained dilute sulfuric acid, prevented the fluids from mixing but allowed the passage of current.[5] The only drawback was that when not in use it needed to be dismantled to stop the chemical reactions and so prevent the metals from being consumed, but for continuous use it was very effective.

Shortly after this, telegraph systems started to appear. Though at first they often used "sand" batteries which were a horizontal version of Volta's "pile" with sand to stop the acid slopping about, they soon converted to using Daniell cells when the service became serious and a reliable battery was needed. In America, they first used the Grove cell, invented in 1839 by an English lawyer with an interest in science.[6] It used zinc and platinum as the electrodes with sulfuric and nitric acids as the electrolytes. Its advantage was that it produced an output of 1.9 volts, but it had the nasty side effect of generating nitrogen dioxide which didn't do the operators any good. Eventually, even they had to use Daniell cells when the traffic increased.

Though there were improvements over the following couple of decades, the next major step was taken by Frenchman Georges-Lionel Leclanché. Unlike many of the others involved in battery development, he was an engineer and became interested in batteries after he had finished his technical education in Paris in 1860. By 1866, he was taking out French patent no 71865 for his cell.[7] It was a major step because it used ammonium chloride as the electrolyte solution, avoiding the use of acids. The cathode was of manganese dioxide with a carbon rod to make the connection, and the anode was zinc.

It gave an output of 1.5 volts though this decreased as the cell aged. Unlike the Daniell cell it could be left assembled as it had quite a long shelf life. Though the output decreased as it was used, there was a degree of recovery if current was not taken for a while. It was thus very suitable for intermittent use. As it was cheap to produce, it was quickly taken up by the telegraph operators and was in service in Belgium in 1867.[8] Leclanché set up a factory to make the cells and their use spread rapidly. By 1868, 20,000 Leclanché-type cells were in use which, in addition to the telegraph, powered alarm systems in coal mines and bell installations, all applications where continuous use was not required.[9]

During the 1870s, numerous attempts were made to improve the Leclanché cell, and though small advances were made, the basic chemistry proved its worth. The main drawback was that the liquid contents meant it could not be moved safely. If it tipped over the electrolyte was spilled. Thus the search was on for a "dry" version.

In the early 1880s, the elements began to come together. The electrolyte was turned into a stiff paste by adding such things as plaster, so that it would not spill. The zinc plate was turned into the container and the carbon electrode was moved to the center. Finally, in 1886 Carl Gassner of Mainz, in Germany, got it all together and patented a new method of construction and produced the first commercially successful dry cell.[10] Progress was now rapid; by 1889 there were at least six well-known makes of dry batteries in circulation, with slightly different detailed arrangements, and they quickly replaced the wet Leclanché cells in their telegraph and doorbell duties.[11]

These were not really "dry" cells, but merely a stabilized version of the Leclanché cell. It isn't really possible to have a truly dry cell, but they were satisfactory for general use as they could be used in any position. Compared to a late twentieth-century battery, they were huge. A number 6 cell, for example, was six inches (15 cm) tall, more than 2 inches (6 cm) in diameter and weighed more than 2 lb. (1 kg).

From the 1890s onward, the battery business was no longer driven by the inventions and by chemistry—it was now a matter of the uses to which they could be put. One of these was for telephones where a battery was needed to power the microphone as well as ring the bell for an incoming call. The early telephones used Leclanché cells, but they rapidly transferred to dry cells. The German Imperial Post Office conducted tests and, finding them superior for telephone use, transferred to Gassner cells in the early 1890s.

The step that changed the dry battery from the preserve of the large institutions running telegraph and telephone operations to a mass market item took place, unsurprisingly, in America. Conrad Hubert was a Russian immigrant who had changed his name from Akiba Horowitz.[12] He tried various trades before setting up a novelty shop under the name American Electrical Novelty and Manufacturing Company in 1898. One of his most popular sales lines were varieties of electric scarf or tie lights. Hubert had a young British man, David Misell, working for him who was a bit of an inventor. That same year Misell came up with a design for a bicycle lamp and also a cylindrical flashlight or torch.[13] These were patented and the patents assigned to the business. They were greatly aided by the fact that the National Carbon Co. had introduced a smaller battery which was the size of the modern "D" cell. While they hadn't invented the flashlight they were the ones to put it on the map.

The devices were called "flashlights" in a rare display of honesty, as with the poor early batteries and carbon filament bulbs they didn't produce light for very long. The name stuck and is still used. Gradually, the batteries improved and in 1903 Hubert patented a variant of the flashlight with a switch that would lock down as well as being pressed to obtain the light.

The real advance came a few years later with the introduction of tungsten filaments for the bulbs. These gave around four times the light for the same power when compared with a carbon filament bulb, so for the same brightness the battery lasted four times as long. With the low voltages required the filament was thick and hence robust. This transformed the flashlight into a valuable device as now the light would last for a reasonable length of time. The transition was shown in the sales of batteries in the US which rose from two million in 1902, to 34 million in 1909, and 67 million in 1912.[14]

Hubert, who was a thorough entrepreneur, set up a factory to make the flashlights and the batteries to go into them. These he sold using the name Ever Ready, and in 1905 he changed his company name to the American Ever Ready Company. The next year the National Carbon Co. bought half the company, and by 1914 had obtained total control. Thus Ever Ready, probably the most famous name in batteries and flashlights, was born.

In 1901, a British company called the American Electrical Novelty and Manufacturing Company Limited was set up to market the Ever Ready novelty products in England though it manufactured the batteries itself. By 1904, it was going its own way and the connection with the American company was severed. Two years later it changed its name to the British Ever Ready Electrical Company Limited (Fig. 10.1).[15]

Fig. 10.1. Ever Ready flashlights, 1903. Source: http://www.gracesguide.co.uk/File:Im19031024
ILN-Ever.jpg

In addition to lighting—this was at a time when very few houses had mains electricity—
the uses now included electric clocks, providing the spark for internal combustion engines,
and powering car horns, as well as many more specialized requirements. With the coming
of World War I the military soon began to depend on batteries to launch artillery and torpe-
does as well as using flashlights in the dark trenches, powering field telephones and radios.

Between the wars, in addition to batteries for flashlights, the other great use was for the
high tension and low tension batteries needed for the booming radio business. Ever Ready
and other battery makers started manufacturing battery-powered radios to boost the consumption of batteries. With the increase in the number of homes connected to the mains,
this market went into decline. The radios were too big to easily carry about and the convenience of not needing to constantly replace the batteries meant that mains radios rapidly
took over.

However, as so often, the coming of war again gave an impetus to battery development.
The forces, particularly the US Army Signal Corps, required large numbers for all sorts of
mobile equipment, particularly radios and mine detectors. There were supply difficulties
due to the need to import the manganese dioxide ore, but in addition the existing unsealed
zinc-carbon cells were proving unsatisfactory, particularly in North Africa and the Pacific.
High humidity caused the zinc to corrode, while high temperatures would quickly discharge the cell even when not in use. The result was that many batteries were flat even
before they had been used or they constantly had to be replaced.

In America, the search was on for a better battery for tropical conditions. One of the
people contacted was an independent inventor, Samuel Ruben. His approach was to build
cells that were airtight, and after many experiments he concluded that the best results were
obtained with a mercury cathode. In late 1942, the Army tested his cells and found that
they resisted ambient effects and stayed fresh on the shelf. In addition, the cell had a "flat"
discharge characteristic with a constant output of 1.34 volts and lasted four to five times
longer than zinc-carbon cells.[16] This was exactly what was required.

The Signal Corps was desperate for the better batteries but, as Ruben pointed out, he was just an inventor. His suggestion was to use the P.R. Mallory company who he had worked with before. Reluctantly, Mallory started to put the cells into production but had many difficulties. The Army—in a hurry, there was a war on—got several other companies, including Ray-O-Vac and the British Ever Ready, to also make them. By 1945, Mallory and Ray-O-Vac were each producing one million cells per day.

In the rush of wartime, the problems of the battery had never really been solved, and considerable further development was necessary to produce a battery for the commercial market. Though some batteries were used in commercial hearing aids in 1946, they were not listed in Malloy's catalogue until 1952. This was just in time for the post-war move to miniaturization. Without these small long-lasting batteries, hearing aids would not have been practicable.

Though these cells were very useful for small items, their higher cost meant that the bulk of the market was still for the standard zinc-carbon cells. The production in America reached two billion cells in 1952. In Britain, Ever Ready, which had obtained a dominant position, rode the boom of the 1950s and 1960s producing batteries for the burgeoning fashion for transistor radios. It produced layer batteries to give the 9 volts required and these were popular until the late 1960s when there were increasing imports of radios from Japan, which used multiple round cells. Also, the growing market for toys and other items that used round cells expanded the business.

By 1962, British Ever Ready was producing more than a billion cells per year, a situation that wouldn't change significantly for the next decade. They had been making steady technical improvements and claimed that from 1950–1972 round cell batteries increased in performance by around 60–90% according to type. At the same time, the storage life was improved by about one hundred per cent. They had also introduced a metal jacket on some round cells to minimize corrosion if leakage occurred.

In 1973, British Ever Ready had such a dominant position, supplying seventy-five percent of the British market, that their activities were investigated by the Mergers and Monopolies Commission.[17] Though they were found not to be abusing their position, and the company was riding high, they seemed to be unaware of the threat they faced. As so often happens, a company with a large investment in a particular technology was reluctant to consider that something could come along to disturb their comfortable position.

The mercury cell performed much better, but it was something like twelve times the price of a zinc-carbon cell. This was driven mostly by the cost of the expensive mercury cathode. The hunt was on for a cell with these advantages, but cheaper to make. An obvious line was to try to make a hybrid of the two. If the expensive mercury could be replaced by the cheaper manganese dioxide of the zinc-carbon cell, and an alkaline electrolyte like the mercury cell was used, then it might have the right characteristics.

In 1949, a Ray-O-Vac engineer, Stanley Herbert, working from their wartime experience of making the early mercury cells, produced a "coin" cell to this formula. Technically it performed well, but they were unable to achieve much market penetration. However, it was enough to attract the attention of their competitors.

Union Carbide, the American Ever Ready, produced a manganese alkaline cell in 1957, while Mallory, again working with Samuel Ruben, brought one out in 1961. Though they had nearly the performance of the mercury cells, their cost was only a little more than three times the equivalent zinc-carbon cell.[18] They also had the environmental advantage of not using toxic mercury. For many higher power applications they worked out cheaper over the life of the cell, but they were still slow to catch on.

Mallory had originally been a supplier of parts to other manufacturers and didn't have the sales structure to deal directly with the consumer market, but it transformed itself in the early 1960s. In 1965, it came up with the name Duracell for the batteries and marketed them under that name. An early stunt was to supply the alkaline cells free of charge to Kodak to put in their cameras. These were sold with a warning that the guarantee would be invalid unless the batteries were replaced by Mallory alkaline cells. This produced an instant market.

Steadily, Duracell's alkaline cells made progress against the entrenched zinc-carbon batteries, particularly to power the new electronic devices like portable cassette players. By the late 1970s, Mallory was gaining a considerable percentage of the market. In 1980, Union Carbide saw the threat to its American Ever Ready brand and launched its own equivalent to the Duracell, which it called Energizer. In the early 1980s, the battery industry grew at around seven percent a year, but for the alkaline segment it was close to double that.[19] The British Ever Ready, sticking to zinc-carbon cells, didn't react, and went into sharp decline, eventually being bought by the same company that owned the American Ever Ready.

By the end of the century the alkaline battery was reigning supreme. The market had very much become global with around 25 billion a year being sold. All other forms of primary batteries, including the zinc-carbons that were still holding on, sold around ten billion.[20] In a little more than a century the dry cell, basically a derivative of the Leclanché cell, had become the mainstay without which so many items, such as the ubiquitous television remote controls, would not have been possible.

Around 1859, Frenchman Gaston Planté invented an entirely different sort of cell – one that could be recharged. He took plates of lead with a textile separator and immersed them in dilute sulfuric acid. When a current was passed through this, lead dioxide was formed at the anode. If this was left and then discharged, and this cycle repeated numerous times, gradually spongy lead and lead dioxide were created in a process called "forming". This increased the capacity of the cell.

Though these cells could produce large currents, they didn't last very long, and it was another 20 years when another Frenchman Camile Fauré found a solution in 1881. His method was to have a skeleton framework of lead with spongy lead and lead dioxide filling the spaces between the ribs.[21] This was a battery which could easily be made, produced a high power for a reasonable time and, crucially, could be recharged from the dynamos that were now available. The lead acid battery, which is still in use, was born.

The large amounts of power produced by these batteries meant that they found use almost immediately for traction vehicles, such as streetcars, trams and electric cars, as well as for lighting train carriages. This was despite their enormous weight. In 1888, a 660 Ah accumulator could weigh as much as 2.5 hundredweight[22] (127 kg), whereas a modern lead acid battery of the same power would only be about a quarter as heavy. The early electricity supply organizations who were generating direct current found them useful to help manage the large peaks in demand in the evenings due to the lighting load (see Chap. 3).

The turn of the century saw two further, though related, inventions: the nickel cadmium and nickel iron, or NiFe, cells. The nickel cadmium came from the work of Waldemar Jungner in Sweden. He also looked at using iron instead of cadmium but opted for the second because it had a greater power for the weight, even though it was more expensive.

In America, the ruthless Edison went for the nickel iron combination and patented it. He was unaware of Junger's work.

The result was that the nickel iron battery, which was a very robust device, was a serious competitor to the lead acid one, whereas the nickel cadmium didn't gain much popularity for half a century. Edison was keen to get his NiFe cells adopted for automobiles, but in the end he largely lost out to the improving lead acid batteries.

The early automobiles used large dry batteries to provide the energy of the spark to their gasoline engines. By the time that tungsten filament lamps were available, automobiles were beginning to use electric headlights instead of acetylene ones. This put a strain on the batteries. However, in 1912 the American Charles Kettering produced the first practical self starter car.[23] (Before that it had been necessary to go to the front of the auto, insert the starting handle and swing on it. If you were lucky, it would start and then you were able to get back in and drive off).

General Motors were keen to fit these electric starters to their automobiles, but it required a far more powerful battery. The obvious candidate was a rechargeable battery, and the lead acid version won the war. The improvements made by the Willard Storage Battery Company in 1915, using rubber for plate separators and for the cases, helped lead acid batteries to really take off, particularly when dynamos were fitted to the vehicles to recharge the battery rather than having to charge it from an external source.

Two versions of the basic design grew up. The starter battery, used in automobiles, could supply a very large current for a short time but was relatively fragile, and the more robust general use battery couldn't supply the high power but lasted longer and could be discharged completely without problems. These were used for traction and static battery systems. With these developments the secondary battery market remained stable for around half a century. Though sales volumes increased there were no marked changes in technology or in their uses.

The nickel cadmium cell (NiCd), invented at the beginning of the century, didn't find widespread applications. However, its superior performance over lead acid batteries meant that it was useful in critical applications, such as for security, starting airplanes and important standby systems.[24] This was particularly so after 1932 when Sabine Schlecht and Hartmut Ackermann in Germany invented the porous sintered pole plate. This provided a larger effective electrode surface area, giving a lower internal impedance and hence a higher current capability.[25]

In 1947, Georg Neumann, the German microphone manufacturer, fine-tuned the chemistry of the nickel-cadmium battery, allowing it to be sealed and its use became widespread.[26] In his version, the gases generated were recombined so there was no loss of electrolyte. This produced a cell which was as convenient as the dry cell in that it could be used in any position, thus opening the way for cordless appliances.

Even then progress was slow and it wasn't until the 1960s, after further improvements, that they started to be used in large numbers in consumer items such as electric razors and toothbrushes. By 1974, manufacturers such as Ever Ready in the UK were making NiCd cells in the same D, C and AA sizes as their dry cells.[27]

Though cordless items such as hedge trimmers started to appear around this time, it wasn't until the 1980s that the market for cordless electric screwdrivers and drills took off. They, of course, depended on NiCd batteries. The rise in consumer electronics was also made possible as many were too power-hungry to satisfactorily use dry batteries. The rise was such that by the end of the century 1.5 billion NiCd batteries were being produced annually, and being used in everything from power tools to mobile phones and computers.[28]

It was now that applications drove research into better rechargeable battery systems. A nickel hydrogen system had been used in spacecraft, but this required a store of high-pressure hydrogen. The nickel metal hydride (NiMH) battery was invented in 1967 as a way of storing the hydrogen in a convenient form. It was patented by Klaus D Beccu working at the Battelle Geneva Research Center.[29] Though it was of interest to Daimler Benz, it still was not satisfactory for general use.

In 1986 Stanford Ovshinsky, founder of Ovonics, patented an improved version.[30] This was licensed to numerous manufacturers and the NiMH battery really took off. It is a more or less drop-in replacement for the NiCd but has a higher capacity. Though more expensive, the NiMH cells are useful where higher energy density (the power produced for a fixed weight) is important. Another factor which drove their acceptance, particularly in Europe, was that there was legislative pressure to remove the use of heavy metals such as cadmium in waste; this accelerated the switch over and expanded their use into items such as mobile phones.

The search was still on for even higher energy densities, and one obvious way was to try to use the lightest metals possible. This led, in 1979, to the American researcher John B. Goodenough, working at Oxford University, perfecting the lithium-ion rechargeable battery. Unfortunately for him, he had to assign the patents to the organization funding his work, the UK Atomic Energy Commission.

The technology was licensed to Sony but it took them until 1991 before they were able to launch Li-ion batteries. These steadily improved until they were superior in performance to NiMHs and some four times better than the performance of lead acids. As they became cheaper than NiMH systems, Li-on found a ready market in digital cameras and laptop computers, which were not really practicable without these advanced batteries.

The Li-ion battery has a drawback in that it has a tendency to become unstable if it is overheated, overcharged or punctured. The thermal runaway that occurs is known in the trade as "venting with flame" or, in common parlance, it catches fire. The battery industry prefers the term "rapid disassembly"! While the smaller types are satisfactory, work still continues to make safe versions suitable for large-scale applications such as in electric cars. (More on this subject in Chap. 15).

The search for the perfect battery continues, driven by the lure of electric cars, but it still seems some way off. Though batteries have steadily improved progress has only been very slow over their 200-year history.

NOTES

1. Kings College London, John Frederic Daniell, Famous King's People available at: http://www.kcl.ac.uk/aboutkings/history/famouspeople/johnfredericdaniell.aspx.
2. DNB, John Frederic Daniell.
3. What-when-how, Daniell, John Frederic (1790–1845) English chemist, meteorologist (scientist), available at: http://what-when-how.com/scientists/daniell-john-frederic-1790-1845-english-chemist-meteorologist-scientist/.
4. *The Times*, January 22, 1825.

5. Corrosion Doctors. John Frederic Daniell (1790–1845), available at: http://www.corrosion-doctors.org/Biographies/DaniellBio.htm.
6. Corrosion Doctors. Sir William Grove (1811–1896), available at: http://www.corrosion-doctors.org/Biographies/GroveBio.htm.
7. Culp B. Georges Leclanché, available at: http://www.chemistryexplained.com/Kr-Ma/Leclanch-Georges.html.
8. *Encylopedia Britannica*, Georges Leclanché.
9. The Monopolies and Mergers Commission: A Report on the Supply of Primary Batteries 1974.
10. Mertens J. The development of the dry battery: Prelude to a mass consumption article (1882–1908). *CENTAURUS*, 42, 109–134, 2000.
11. Solar navigator, battery development and history, available at: http://www.solarnavigator.net/batteries.htm.
12. NNDB, Conrad Hubert, available at: http://www.nndb.com/people/439/000169929/.
13. US Patents 603112 and 617592.
14. Mertens for 1902 and 1912. His figure for 1909 seems high and so the figure is taken from: Morehouse C.K., Glicksman R. and Lozier G.S. Batteries. *Proceedings of the IRE*, August 1958.
15. Tweedie A. Ever Ready Co (Great Britain), available at: http://www.gracesguide.co.uk/Ever_Ready_Co_(Great_Britain).
16. Hinz E.S. Portable power: Inventor Samuel Ruben and the birth of Duracell. *Technology and Culture*, 50, Jan 2009.
17. The Monopolies and Mergers Commission, A Report on the Supply of Primary Batteries 1974.
18. Hinz.
19. Energizer Holdings, Inc., Company History, available at: http://www.fundinguniverse.com/company-histories/energizer-holdings-inc-history/.
20. Calculated from I. Buchmann, Battery statistics, available at: http://batteryuniversity.com/learn/article/battery_statistics.
21. Witte A. The Automobile Storage Battery its Care and Repair, 1922, Ch. 2, available at: http://www.powerstream.com/1922/battery_1922_WITTE/battery_WITTE.htm#toc.
22. Derry T.K. & Williams T.I. *A Short History of Technology*. Oxford. 1960, p. 612.
23. Lawson B. Electropaedia, History of batteries, available at: http://www.mpoweruk.com/history.htm.
24. Green A. Alcad history: The story of 75 years of NiCd battery manufacture in Redditch, available at: http://www.alcadhistory.org.uk/technology.html.
25. Lawson B, Electropaedia.
26. Neumann Berlin, Georg Neumann – An inventor and his life's work, available at: http://www.neumann.com/?lang=en&id=about_us_history_part_1.
27. The Monopolies and Mergers Commission: A Report on the Supply of Primary Batteries 1974.
28. Wikipedia, Nickel-cadmium battery, available at: https://en.wikipedia.org/wiki/Nickel%E2%80%93cadmium_battery.
29. Lawson B. Electropaedia.
30. *The Economist*, In search of the perfect battery. March 6, 2008.

11

A Good Investment: Electricity Grids

We are witnessing a stage in the miracle of electricity... We look back on the countless centuries which have made this island the most beautiful in the world. Then came man with his industrial revolution, and we have covered parts of the country with dirt, smoke, and poverty... Now we have called in science to our aid, and electricity is going to be its handmaiden.

Stanley Baldwin at the opening of Stourport Electricity Station, 1927

From the chaos caused by numerous small local electricity generators, a few organizations such as NESCo in north-east England, Commonwealth Edison in Chicago, and Rheinische Westfalisches Elektrizitalswerke (RWE) in the Ruhr and the Berlin system in Germany stood out as building integrated networks.[1] In all the developed countries a mixture of public and private ownership had grown up which often militated against cooperation. Though the lessons of the technical advantages of these networks were understood, it was the organization of the industry which stood in the way of further progress.

Only in Germany, where the organizations were often at least part owned by the local authorities, was the division avoided, but was replaced by a parochialism that set them against connections with their neighbors and hence building larger networks. In America, they often suffered from both problems as a layer of state regulation had been added to try to control what was increasingly being seen as another utility.[2]

However, it was in America that a plan was begun to start a network to cover a larger area. The "Superpower" network was proposed by William S. Murray and his associates in a report of 1921. He advocated a large-scale study to measure the benefits of an electricity network. The area proposed was the industrial north-east, in a region running from Boston to Washington, D.C., which was called the Superpower zone.

The plan was to retain only the most efficient generating plants and build new ones to supply the network, with considerable savings. Thousands of miles of high voltage lines were to be built and, despite investment of around $90 million a year for five years and $48 million for the next five, it was expected to save some 40% of the projected costs of supplying this area without the network.

© Springer International Publishing AG 2018
J.B. Williams, *The Electric Century*, Springer Praxis Books, DOI 10.1007/978-3-319-51155-9_11

Despite the setting up of a Superpower Supervisory Board and the involvement of Secretary of Commerce Herbert Hoover, the conflicting interests of the various organizations could not be reconciled. It wasn't helped by Murray taking a position strongly against governmental involvement in electricity utilities, when it was probably the only organization that could enforce the scheme. He was advocating unified ownership of the utilities, but that cut across the desire to control monopolies.

Though the technical details were well worked out and the advantages clear, the scheme foundered on the conflicting interests with no overarching organization to resolve the problems. Despite this, another scheme, "Giant Power", was proposed, this time in Pennsylvania. It was a similar scheme in many ways but was more adventurous in the technical proposals it advocated.

The industry was to be divided into three parts: generation, transmission and distribution. This required very considerable reorganization and the existing utilities were united in their hatred of it. Murray even went so far as to describe the scheme as "communistic". Despite a long battle it was never passed by the state legislature of Pennsylvania. Thus neither of these schemes, with clear technical advantages, saw the light of day. The country carried on building the system piecemeal.

It was in Britain, which was lagging in the race to electrify, that a solution to these problems was eventually found. It was war, as usual, that starkly exposed the weaknesses of the British electricity supply system. Though demand fell at first, as people economized, it soon started to rise. A particular problem was that the high-grade Swedish iron ore supplies had been cut off, and it was necessary to use electric arc furnaces to process the local ores.[3]

The large numbers of young men entering the services led to manpower shortages in industry and particularly in the coal mines. This transformed the dull subject of "efficiency of generation" into one of national importance. It was vital to reduce the amount of coal being consumed, and as so much electricity was being used for war work it couldn't simply be switched off. It was essential to get more power from every ton of coal.

Early in 1916, Ernest T. Williams, a leading member of the Institution of Electrical Engineers, introduced some proposals for a new Electricity Board to tackle the problem. He felt that it was not so much one of technology or of finance, but of organization.[4] Basically the problem was parochialism. The ill-advised Electricity Acts had bred this mentality and, with few exceptions, the local providers, whether they be municipal or company, fiercely defended their patches and were reluctant to cooperate. This was despite the experience of NESCo in the north-east, and numerous other examples, where increasing size and area covered produced a better and cheaper supply for consumers. Now with a war on, it was a matter of national importance.

The government set up a Department of Electric Power Supply, wisely choosing William McLellan, Charles Merz's partner, as its director, though in 1917 this role passed to Arnold Gridley. Though some progress was made it tended to degenerate into an arbiter of wartime priorities. Probably the greatest success was to persuade a number of suppliers in the Manchester area to interconnect their systems which led to considerable savings of capital costs and coal.

When there is a contentious matter, governments always reach for a committee, and electric power was no exception. Such was its importance here that there were three committees to examine different aspects of the subject. They were basically in agreement, so the government had yet another committee. It didn't look as though this was going anywhere.

However, the final report from the Birchenough committee was even more radical than the others. It came to the conclusion that "any efficient system for the development of electrical generation and main line distribution in the United Kingdom must not only be a national system, but a single unified system under State regulation, in the financing of which the State would participate on a large scale. It should, however, be framed and administered throughout upon an entirely commercial basis, and not in any sense on Civil Service lines".[5]

The report went even further being "strongly in favour of nationalization of generating plant" and that "the State should control all main transmission lines". The Electricity Board that they proposed to control this might leave the distribution in the hands of the existing agencies, "where such agencies were efficient and progressive". Some of this was unsurprising when two of the committee members were Ernest Bevin and J.H. Thomas, prominent Trade Unionists leaders, and Thomas was also a Labour MP. However, many of the others were leading businessmen, and it just showed how attitudes had changed due to the government involvement in many industrial matters during the war.

There was fairly general agreement that the present position was a mess. After all, there were some seventy authorities providing electricity just for Greater London, owning over seventy generating stations, with at least fifty different types of system, ten different frequencies, and twenty-four different voltages.[6] That didn't suggest that they were going to agree about the solution.

Though Lloyd George was still Prime Minister after the November 1918 election, he presided over a coalition government containing a large number of Conservatives. There was no way they were going to pass such a socialist measure, quite apart from the opposition of vested interests such as electricity undertakers and municipalities.

For pure constancy of opposition the prize must go to the new Unionist MP for Hampstead, George Balfour. He was an engineer by training and had set up a consultancy, with his colleague Andrew Beatty, called Balfour Beatty. They had been involved in many tramways and lighting projects, but particularly with the management of a number of Midlands electricity undertakings. It was a clear case of vested interest. Balfour was a self-made man who strongly supported private ownership and had a great distaste for the liberal policies which favored state control.

The net result was that the Act came out in emasculated form. It set up Electricity Commissioners to try to bring about reorganization of the industry, but though they were to try to bring the undertakers together to set up Joint Electricity Authorities (JEAs) in the various regions of the country, they had no powers to enforce this. The tide was now running against the forms of state intervention that had been used successfully in the war.

The chairman of the Electricity Commissioners was Sir John Snell, and he and his team worked hard to define what needed to be done and delineate areas for the JEAs, but they constantly ran into the old problems of parochialism. Despite this, a small amount of piecemeal interconnection took place, and some progress was made. In North Wales and in the West Midlands, where a company and several municipalities cooperated, JEAs were set up to bring bulk supply generation to their areas.[7]

Generally, the committee met obstruction particularly where anything involving George Balfour was concerned, and as he had the management of the extensive Power Securities Corporation this was a major problem. It became increasingly clear that voluntary arrangements weren't going to work.

There was a further problem in that though the frequency of the supply had been defined as 50 cycles (Hertz (Hz) in modern parlance), large parts of the country used up to seventeen different frequencies, varying from 25 to 100 Hz. The largest problem lay in what had been the most successful interconnected system, namely NESCo in the North East, where Charles Merz had chosen 40 Hz. Gradually, most people who were involved in this process (apart from George Balfour), realized that a more drastic solution was necessary.

The 1920s were a time of turmoil in many spheres, not least in politics. By 1924, Lloyd George was out, and a Conservative government under Bonar Law and later Baldwin had been and gone. A minority Labour government was now in power. They considered restoring the compulsory powers that had been removed from the 1919 Act and became interested in the frequency standardization problem. Though they were prepared to spend money on this, they needed to know the figures, and commissioned Merz & McLellan to prepare a report.

Before they received the report the government was out and Baldwin's Conservatives were back. However, Stanley Baldwin, despite a rather "arts" background, had experience of business in his father's iron-making concern and an interest in science that led to him being made a member of the Royal Society.[8] He understood the importance of electricity and the need to reorganize the industry.

One of his first acts was to get Lord Weir to set up another committee to investigate what should be done about the electricity industry. It set to work in January 1925 and started with the advantages of the five years' experience of the Electricity Commissioners and also the report on frequency standardization commissioned by the Labour government. They were critical of the "voluntary" aspect of the previous Act and said that "the policy of persuasion can only be written down as a failure".[9] They were concerned that the country was falling ever further behind. Now even NESCo's network had been overtaken by others abroad, particularly RWE in the Ruhr in Germany.

The committee were beginning to think in terms of a national rather than the Electricity Commissioners' regional interconnection systems, and so commissioned three reports on possible schemes, their costs and implications. For these they chose Sir John Snell of the Electricity Commissioners, Charles Merz, and Sir John Kennedy, a partner in the other leading firm of consultants, Kennedy and Donkin. The reports were remarkably similar in detail.

The recommendations were clear in their technical proposals. A national "Gridiron" of interconnections should be built at a cost of $100 million (£25 million).[10] All the non-standard frequencies should be converted to the standard 50 Hz, which would cost a further $42 (£10.5 million). Without this the full benefits of the scheme could not be realized. They projected that by 1940 the present average price of 4 cents (2 old pence) per kWh would fall to 3 cents without the scheme, and 2 cents with it. As the projected amount of electricity used would have risen from the current 125 kWh per head of population to 500 kWh, the savings would be about $180 million (£44 million) a year, or an astonishing one percent of GDP. On these figures the scheme was, in modern parlance, a no-brainer.

The reasons for the savings were the same as had been proposed by the American schemes. The new body would have the power to select the most efficient stations and not take supplies from the most inefficient ones. Thus the average efficiency of generation would rise. Also the interconnection would, as NESCo had found out, improve the average load factor, making an enormous difference to the economics. Thirdly, in an industry so dominated by the capital cost of plant, the "Grid" would interlink everyone and so much less spare capacity was required overall.

There is little doubt that Weir overstated the case, but this was to make it so simple and straightforward that the politicians would see the advantages and accept it and implement it. One example of the gloss they put on things was to compare the amount of electricity used per head of population with other selected countries or places where it was much higher, such as Switzerland. As their critics pointed out, this was not reasonable as Switzerland had no coal and abundant possibilities for hydroelectric generation.

The clever part of the proposals was the politics. The fundamental problem was how to gain control of the situation to force the necessary changes on a reluctant industry without nationalizing the lot and running into all the objections of confiscating assets that the companies had spent large sums of money building up. A Conservative party was never going to accept that.

The solution was to propose a new body, the Central Electricity Board, which though state-owned could raise money by commercial borrowing. It would build the Grid. Then it would buy electricity from any efficient stations on its approved list and sell it to the distributors. This way it could finance itself on the margin. The cost of the frequency standardization would be recouped by a levy on the whole industry. Thus it had a grip of the whole situation, but the generating stations would remain in the ownership of the original providers. It wouldn't be confiscating any assets, just providing an economic incentive to efficiency.

This was the art of the possible in the engineering, but the art of what you could get away with in the politics. The report was delivered in a remarkably short time in May 1925. The government went for it, but with such great caution that it took until March the following year before the Bill was introduced into Parliament. Only then was the report published.[11] It was one of the most important Bills of the year.

Of course, it ran into opposition, but most of the more obvious grounds for objection had already been neutralized. George Balfour reappeared to try to continue his wrecking, but with the failure of the voluntary scheme in front of them people were less inclined to listen to him. The government had other local difficulties to deal with, not least of which was a general strike, but they pressed on.

The Electricity Act 1926 just made it on to the statute book before Christmas with the main planks of the scheme still in place. The Central Electricity Board (CEB) was set up in March and went straight to work. It divided the country into nine areas (there was never a plan for the north of Scotland) and defined schemes for each of these. They were so arranged that they could all be connected to form a national Grid. They were also to standardize the voltage and frequency.

The advantage of the separate schemes was that some areas could be agreed fairly quickly and work start straight away, while the more difficult problems could be tackled later. The first scheme was agreed within a few months of the start and then they came thick and fast throughout 1928, 1929 and 1930. The only one which trailed was South Scotland when it was decided to implement a number of hydroelectric schemes.

An early decision was to use overhead power lines. This wasn't as clear cut as it seems now. There was not much experience in the country, and many thought the risks too great, particularly as the need to minimize the losses pointed to using the highest voltage possible. The decision was to use 132 kV. This was courageous as there were no lines of this voltage in the country, but some of the suppliers had experience of building them abroad. The Board went into this carefully and examined the experience with systems around the world. There were, after all, some advantages in being a little behind.

The next task was to obtain the "wayleaves", that is, the permissions from around 22,000 landowners to run the cables over their land, and plant the towers to hold them. This was a tremendous undertaking for more than 4000 miles of lines, and about 26,000 pylons, as the towers became known.[12] Obviously this was a fairly delicate task, as not everyone was too keen to have a 70–80-ft high pylon on their land. Often the projected routes had to be modified as they went along to avoid "obstructions". These were difficult people who just couldn't be persuaded.

By the middle of 1930, more than a 1000 miles of wayleaves had been obtained. Within another 18 months this was up to 3500 miles.[13] The construction of the pylons followed this pattern with about a year's delay. This was a tremendous effort as even the smallest pylons contained some three tons of steel. The largest were on the river crossings, with those on the Thames 487 ft. high and weighing 290 tons (Fig. 11.1). Altogether the scheme,

Fig. 11.1 Grid cables being taken across the Thames to join the schemes in Essex and Kent. One of the enormous 487 ft. towers can be seen 3000 ft. away on the other side of the river. Source: "The Times", September 17, 1932.

including switching and transforming stations, consumed around 150,000 tons of steel and 15,000 tons of aluminum in the cables.[14]

While the whole scheme had begun in relatively good times in 1927, by the time that this main construction effort was required the world was in the Great Depression. The timing was thus doubly fortunate. Firstly, the steelmaking and processing industries were slack so the parts for the pylons could be constructed without delay, speeding up the progress. Secondly, it involved just those industries where unemployment, both actual and potential, was at its worst. In all, both directly and indirectly, it was estimated that the Grid employed some 200,000 people in its construction.[15]

As can be seen from Fig. 11.2, the dip in GDP in the years of the depression is not very noticeable. The peak of the spending was in 1931 and 1932 which, fortuitously, was exactly when it was needed to minimize the slump. Undoubtedly this heavy spending in industries which were suffering contributed to the minimizing of the depression in the UK. It was noticeable that the reduction in activity was markedly less than in other countries, particularly France, Germany and the USA where the drop in GDP was five to six times as great.[16]

On September 5, 1933 the last pylon, which was in the New Forest in the south of England, was completed. This was heralded as the completion of the Grid[17] although this wasn't entirely the case as there were still cables to be slung and switchgear and transformers to

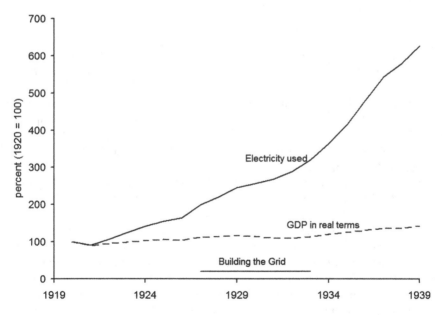

Fig. 11.2 UK interwar electricity consumption compared with the change in GDP, 1919–1939. Source: Author.[18]

be installed. An even greater problem was the frequency standardization. This was a larger task than had been first thought because it was trying to hit a moving target. Between the first estimates in 1924 and the actual implementation a few years later, of necessity, a considerable amount of non-standard plant had been installed. All this had to be changed.

It wasn't only the turbo-alternator generating sets, though there were 160 of these, but some 100,000 motors in the hands of the 40,000 customers.[19] Half of these were in the North East in NESCo's 40 Hz network. The task was so huge that serious consideration had been given to leaving a couple of areas outside the national scheme. What changed matters was that the government offered an Unemployment Assistance Grant. This meant that some 3000 men were employed on the conversion work in the North East, one of the worst-hit areas.[20]

Steadily, from the end of 1932 the areas of the scheme were brought into full operation with the CEB buying the electricity from the generators it chose on its approved list, and selling it to the distributors. By the end of 1934 all but two of the areas were fully in operation. The stragglers were the South of Scotland where the hydroelectric schemes were still being built, and the North East, where the tremendous task of all the conversion took until the middle of 1938.

At first the regions operated independently, but during a day in November 1936 six Grid areas were connected together to form a single system with power supplied from 112 stations controlled by the CEB with another 45 stations linked to it.[21] This was the largest number of stations that had ever been connected to a system at the same time.

A year later, the whole Grid was operated as a single unit for a few hours. This was the prelude to the setting up a National Control Center and the Grid being normally run as one interconnected system. Its true value was shown just before Christmas 1938 when increased demand for rearmament, and delays in bringing new generating plant into service, coincided with an abnormally cold week. There was a considerable shortage of power in the south, but this could be made good by importing power from the north. The system held the high peak load without disconnections.

What had been achieved was the bringing together of the entire generation system, and Britain was the first country to do this. In one bound the industry had caught up and in many cases passed most other countries. The acceleration in the use of electricity can clearly be seen from Fig. 11.3, bringing the country up to around the 500 kWh per person figure that had been projected in the Weir report.

The real question was whether the projected savings were true. The cost of the Grid had come out at $107 million (£26.7 million), which was only a little higher than the estimate, but the frequency standardization had ballooned to $70 million (£17.3 million).[22] This increase was largely due to the amount of additional plant that had to be converted. So the bill was somewhat higher than had been projected, at $180 million (£44 million). Some of this could be attributed to plant that needed replacing, and some of it was covered by Unemployment Grants, and bringing people into work.

The interconnection reduced the amount of spare plant in the system and by early 1936 it was estimated that more than $44 million (£11 million) had already been saved.[23] By the following year this had risen to $88 million (£22 million).[24] Using the most efficient stations reduced the average costs of fuel and other operating costs, producing savings that could be passed on.

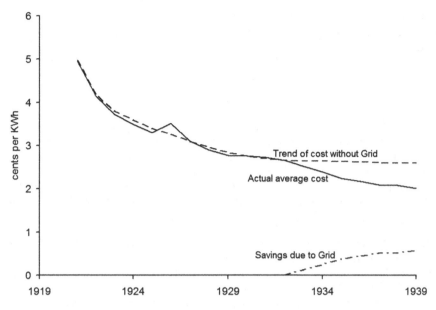

Fig. 11.3 Estimated savings due to the Grid, 1919–1939. Source: Author.[25]

Of course, the true savings can never be known, but Fig. 11.3 attempts to get an idea of them. As the grid came into operation there was a sharp fall in the average price of electricity, despite the fact that the price of coal was rising. Taking this into account, a likely curve that the price might have taken is shown dotted. Plotted as a continuous line, is the actual price. Below them is the difference, or theoretical saving. The price had fallen to around the 2 cents per unit projected by Weir, but without the Grid it would probably have been about 2.6 cents rather than their 3 cents.

Despite that, the saving in 1939 was around $108 million (£27 million) and total to that date around $400 million (£100 million). While this is somewhat short of the Weir report projections, the savings were huge. In a few short years, the country had got back all of its investment and was in clear profit. There are not many investments were that achievable. Thus, the confidence of the scheme's promoters was not misplaced.

However, the benefits were much greater than the narrow financial ones. The reduction in price had spurred greater use which was to give industry advantages in so many ways. Households had taken to electricity in a big way as will be seen in the following chapter. The standardization of voltage and frequency meant that standard appliances could be sold throughout the country.

An unforeseen advantage was that with another war imminent, the country was in a much stronger position with an efficient generating system, and a resilient network that would be a great strategic asset once the bombs began to fall. It was to take other countries many more years to introduce their own Grid systems, obviously following the example. In America, there is a school of thought that considers the problems were never solved and that the great blackouts in 1965 and 2003 were a result of imperfect networks.

When we see those pylons still marching across our countryside, we should remember, Weir, Merz, Williamson, Duncan, and all those others, whose vision made it possible.

NOTES

1. Millward R. Business and government in electricity network integration in Western Europe, c. 1900–1950. *Business History*, 48:4, 479–500, October 2006.
2. Hausman W.J. and Neufeld J.L. The economics of electricity networks and the evolution of the U.S. electric utility industry, 1882–1935. *Business and Economic History*, 2, 2004.
3. This chapter leans heavily on Hannah, *Electricity Before Nationalisation*, Chapters 2–4.
4. Williams E.T. The electricity supply of Great Britain. *Journal of the Institution of Electrical Engineers*, 54:260, 581, June 1916.
5. *The Times*, April 4, 1919.
6. The Lord Chancellor, Lord Birkenhead, speaking in the debate on the Electricity Bill, Hansard *HL Deb 03 December 1919 vol 37 cc575–98.*
7. Sharman F. with Parker B. The story of electricity supply in the Wolverhampton area, available at: http://www.localhistory.scit.wlv.ac.uk/articles/electricity/history3.htm.
8. Baldwin studied history at Cambridge. Two of his uncles were the painters Edward Burne Jones and Sir Edward Poynter, while his cousin was the writer Rudyard Kipling; Obituary: Irvine J.C. Earl Baldwin of Bewdley, K.G. 1867–1947, *Obit. Not. Fell. R. Soc.* November 1, 1948.
9. Quoted in Hannah, p. 91.
10. For comparison this would have been over £1 billion in 2000.
11. *The Times*, March 13, 1926.
12. Pylon is the monumental gateway of an Egyptian temple. The connection with the electricity towers is not obvious.
13. Wright J. Inaugral address as President of the Institution of Electrical Engineers. *Journal of the IEE*, 6: 517, 1, January 1940.
14. *The Times*, June 11, 1932.
15. *The Times*, September 5, 1933.
16. Price C. Depression and recovery. In F. Carnevali & J-M.Strange (eds), *20th Century Britain*, p.153; Maddison A. *Dynamic Forces in Capitalist Development*, p. 55.
17. *The Times*, September 5, 1933.
18. Electricity use calculated from Dept Energy and Climate Change, Historical electricity data: 1920 to 2010; Electricity generated and supplied, based on Department of Energy and Climate Change, Digest of UK Energy Statistics (DUKES); Real GDP data derived from the Bank of England, Data Annex to the 2010 Q4 Quarterly Bulletin article The UK recession in context— what do three centuries of data tell us?
19. Wright J. Inaugral address as President of the Institution of Electrical Engineers. *Journal of the IEE*, 86: 517, 1, January 1940.
20. *The Times*, April 7, 1933.
21. Electricity Council, Electricity Supply in the UK: A chronology, c. 1987.
22. *The Guardian*, April 12, 1934; Wright, p. 4.
23. *The Times*, April 1, 1936.
24. Hannah, p. 128.
25. Actual average prices from Hannah, p. 430, the trend cost, and hence savings, are the author's estimates.

12

Willing Servants: The Growth of Appliances in the 1930s

We have learnt that electricity is before all things the greatest labor saver, and a little investigation shows that in the application of electricity to the requirements of the home we have open to us a field almost unimaginably wide.

Sebastian Ziani de Ferranti

The interwar years, particularly the 1930s, are usually thought to have been ones of depression. While it is true that in the UK unemployment leaped during the 1921 recession and didn't come down again until World War II, it doesn't tell the whole story.[1] Even in America, where the effects of the 1929 stock market crash rippled through the economy, still the majority of the population was in work. However, the problems were great enough to temporarily halt the rise of electrification of homes as can be seen in Fig. 12.1.

Problems in Germany were even more severe as the country had barely recovered from the hyperinflation of the 1920s when the Great Depression struck. Progress largely halted until the 1930s when rearmament boosted the economy back to some sort of normality. However, with the regime at the time they were reluctant to say how well or otherwise they were really doing, and so data are scarce.

In Britain the pain of the collapse of world trade in the early 1930s was felt particularly in the exporting industries, and hence in the places where they were concentrated. The most seriously affected was shipbuilding, and this had its home along the banks of the rivers Tyne in north-west England and Clyde in Scotland.

As can be seen from Fig. 11.2 in the previous chapter, while GDP only showed modest growth and that towards the end of the period, electricity consumption roared ahead reaching more than six times its 1920 value. This disguises the regional variations. For example, in 1931 while the electricity consumption of the whole country rose by 545 GWh (around 6%) that of north-east England and central Scotland fell by about 138 GWh. (a fall of 6–7%).[2] Even by 1938 the average domestic electricity consumption in the North East was only 45% of that in the South East.[3]

© Springer International Publishing AG 2018
J.B. Williams, *The Electric Century*, Springer Praxis Books, DOI 10.1007/978-3-319-51155-9_12

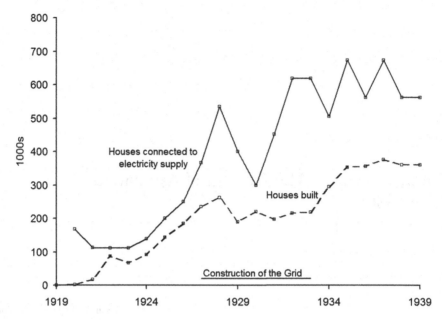

Fig. 12.1 Numbers of new houses built each year in Britain 1919–1939, and the number connected to the electricity supply. Source: Author.[4]

While those out of work suffered terribly, for those lucky enough to have a job (the vast majority) it is often said that it was a good time. Between the wars some four million new houses were built—many of them semi-detached houses built stretching out of towns in what is known as ribbon development. The number constructed each year steadily rose during the 1920s, but in the 1930s, despite the Depression, it never fell below about 200,000 a year. Of course, these were predominately in central and southern England, and far fewer were built in the worst-hit areas.

Nearly all new houses were wired for electricity, but increasingly older properties were having it installed. In the early and mid 1920s it can be seen from Fig. 12.1 that it was mostly the new house building that drove the increase in houses connected. Around 1927 and 1928 there was a sudden rush of exuberance. This was just at the point where the Electricity Act had been passed and the hype for what the prospective electricity grid would do to electricity prices was at its height.

Once the Depression started to bite and reality taking over, the numbers started to dip. Subsequently, with the real benefits of the Grid beginning to work through, and assisted wiring schemes becoming available, the numbers climbed again. From then on they were always in the region of 500,000 to 600,000 per year, which was more than the total of all those connected before 1919. This meant that houses were being wired up at the rate of 2000 a day or 10,000 a working week. Thus the proportion of houses wired for electricity rose from around 6% at the end of World War I, and despite a considerable increase in the

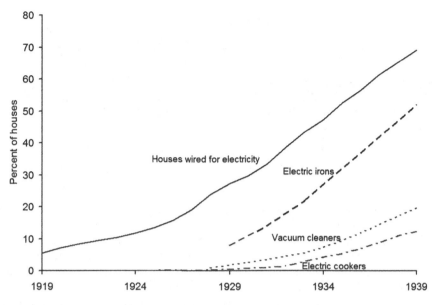

Fig. 12.2 Percentage of British houses wired for electricity and estimates of the proportion of houses that contained various appliances, 1919–1939. Source: Author.[5]

number of households, to around two-thirds by the beginning of World War II. It was an extraordinary transformation (Fig. 12.2).

It had long been known that electric current flowing through resistance wire produced heat, but it took until the 1880s and 1890s before serious attempts were made to incorporate this concept into practical appliances. The Canadian Thomas Ahearn may have invented the first electric oven in 1882, but it doesn't seem to have been put into service until a decade later.[6] Certainly, an electric stove was exhibited at the Chicago World Fair in 1893.

Obviously, a bare wire is dangerous and it wasn't until H.J. (Herbert John) Dowsing managed to sandwich the heating element between metal plates that practical hotplates became available.[7] By 1892, he was giving demonstrations of cooking by electricity, and was going around the country giving talks on the subject.[8] It seems strange now but people needed to be convinced that it was possible to cook without a flame. By the turn of the century Dowsing was producing heaters using a form of electric lamp (still a carbon filament at this stage) which was run at a lower temperature so that it produced only a slight glow and nearly all the energy was turned into heat. This nicely overcame the problems of the resistance wire at the time which tended to rust.

In 1906, American Albert L Marsh invented Nichrome wire, which was a more flexible and satisfactory solution for the resistance wire, and a vast array of appliances started to appear.[9] In addition to cookers there were fires, kettles, grillers, water heaters, saucepans, toasters and irons (Fig. 12.3).

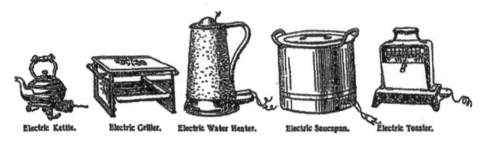

Electric Kettle.   Electric Griller.   Electric Water Heater.   Electric Saucepan.   Electric Toaster.

Fig. 12.3 A range of small appliances dating from 1913. These were in addition to cookers and irons. Source: Electricity in the Home and Office, *The Times*, December 1, 1913.

While cookers, fires, kettles, toasters and irons are still with us, the others have largely disappeared. This is probably because they were used by the middle classes (who could afford the expensive electricity) in their separate "breakfast rooms". Though a vast array of items was available the numbers sold were not great due to the limited numbers of houses with electricity. Even in those that did have it, most were only wired for lighting. For a cooker, a higher-power connection was needed and this was a big obstacle. Few houses had power sockets and so the items were usually designed to consume a low enough power to be run from a light fitting.

Charles Reginald Belling was a Cornishman, but left his home county in the pursuit of technical training.[10] In 1912, at the age of 28, he set up Belling and Co. with his friend Charles Arnold, who went on to found MK Electric after the war. Belling's idea was to wind coils of resistance wire on the front face of fireclay, hence producing elements that could achieve a high temperature. With the addition of a reflector behind this made a useful electric fire. These were a considerable improvement on the Bastian heaters which used the wire inside a quartz tube.

After World War I was over there were opportunities for an enterprising manufacturer, and Charles Belling was certainly that. He quickly introduced his "Modernette" cooker and by 1921 had a large range of fires, kettles boilers, and even a tea urn, whose design doesn't seem to have altered since. In 1929, he introduced the first "Baby Belling" which became the standard fitting for all bed-sits. By the mid- to late-1930s there was a range of cookers with features such as glass doors and in the cabinet style that is still with us.

The success of Belling and the other companies was due to a number of factors. Obviously, the growth of the electricity supply was essential, but there were a number of interrelated effects which defined the period. Particularly in Britain, there was much talk of "the servant problem" after World War I. It is commonly assumed that this was due to declining numbers willing to do the work, but it isn't as simple as that. The numbers of domestic servants hardly altered between the censuses of 1911, 1921, and 1931, though no doubt they ebbed and flowed between these dates.[11] There was no further census until 1951 when the numbers had dropped considerably, so it is difficult to really know when the numbers began to fall.

So why was there such a servant problem almost from the end of the war? Conventional economics would suggest that it was the demand that had increased, and in the absence of data this seems likely, with a steady growth in the number of households due to population growth and a decrease in family size. With the expansion of the middle class more families would aspire to that life and having servants.

Another factor was the attitudes of the potential employees changing faster than those of those who would employ them. Servants were less and less inclined to accept the restrictions which had been traditionally put on them, such as their living conditions, long and awkward hours of work, restrictions on when they took time away from the house, and whether they could have "followers". The result was that there were constant falling outs and servants leaving. This "churn", in modern parlance, was at least part of the perceived problem. The greater number of opportunities for young women outside domestic service undoubtedly contributed to the instability.

Even as early as 1924 it was accepted that "practically every housewife nowadays has to do part of her own domestic work".[12] Electrical appliances were seen as "willing servants" who would always be available and not have off days and threaten to leave. This could be quite attractive to the housewife who had just lost a servant for the umpteenth time. Sometimes it was the possession of an appliance, hence reducing the work, which could lead to servants being attracted to or staying in a particular position.

In America, with a very different society, servants were an even greater problem and so there was more pressure on the housewife. This is undoubtedly one of the factors that drove electrification in the home, and particularly the take-up of electrical appliances, a great deal faster than in other countries, as can be seen by comparing Figs. 12.2 and 12.4. Other European countries were more in line with Britain.

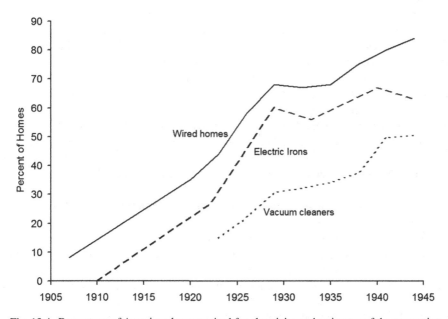

Fig. 12.4 Percentage of American houses wired for electricity and estimates of the proportion of houses that contained various appliances. The curve for electric irons is speculative, due to sparse data, but is in the generally correct region. Source: Author.[13]

New houses at this time were constructed on very different patterns than previously to minimize the need for servants and make them more convenient to service. An important factor in this was the ability to bring heat and light to all parts of the house at the flick of a switch. These trends were interrelated and fed off each other.

What was clear was that, if electricity could be supplied to all houses at low cost, there would be a very considerable demand for "labor saving" devices. As the 1920s wore on this was one of the pressures for radical reform of the supply industry which led in Britain to the creation of the Grid and other attempts elsewhere. Coupled with this were the standardization of voltage and frequency, and the introduction of standardized plugs and sockets. It was now possible for manufacturers to mass produce appliances, freed from the need to make many different variations.

One of the factors holding things back was the high price of electricity. Some of this was due to the complicated tariffs used by many of the suppliers, but also with the general level of the charges. This there was the potential to create a virtuous circle where lower prices would boost demand which in turn would allow prices to be reduced further. The question was: exactly how to get this rolling?

The answer was a campaign to teach women about the advantages of electricity in the home, and this began by educating them to become demonstrators of electrical equipment. Many women were still afraid of electricity and this could only be overcome by a large and sustained program to show them its benefits, and that it was quite safe and gave substantial advantages over other methods. No more filling and lighting smelly oil lamps, no more carrying coal and the dirt associated with this. There was also a suggestion of the luxury of having a switch by the bed so the electric fire could be turned on to warm the bedroom before you got up.

These campaigns seem to have had an effect, as can be seen by the considerable increase in houses connected to the electricity supply in 1926–1929 in Figs. 12.2 and 12.4. Around this time appliances started to be sold in appreciable numbers. Figures 12.2 and 12.4 show speculative curves for electric irons, but fires were probably similar though there are no reliable figures for the number in use.[14]

The manufacturers of appliances were also beginning to consider the desires of the consumers much more, rather than just the technical aspects of the product. This was particularly so where the items were taking over from some existing product, as the customer needed to be led away gently to something that was at least partially familiar. A classic example of this was Herbert Henry Berry's imitation fire.[15] This used an electric lamp to provide light under a layer of material arranged to look like a glowing coal fire. Below this, but above the lamp, was a small fan turned by its heat which gave a flickering effect. The real heat came from separate electric elements. This was the forerunner of the popular Berry Magicoal fire, variants of which are still with us.

Electric irons had been around since the 1890s, replacing smoothing irons heated on the stove and such evil devices as the gas iron with flames coming out of its sides. The electric iron was just plugged in until it was hot enough, and then unplugged or switched off. Thus the user had to control its heat in much the same way as the traditional iron which was put back onto the stove to reheat once it became too cool. The great advantages were that it was clean and much lighter.

The first successful iron was produced by Californian Earl H. Richardson in 1905. He arranged the heating elements so that the point of the iron was as hot as the center— hence the brand name Hotpoint which is still used. However, it took until the 1930s before irons with thermostatic controls started to appear. This seems an extraordinarily long time, but although it is much easier it requires a change in thinking about how an iron is used. It is a clear example of how development was often much slower than might be expected, and what now seem obvious improvements took a long time to be adopted.

Electric stoves or cookers had been around since the 1890s, but it was only after World War I that compact units, suitable for the smaller home, particularly in Britain with its tiny kitchens, started to appear. Charles Belling's "Modernette" was a good example. Though there were some improvements in the use of pressed metal instead of castings, and the introduction of vitreous enamel coatings, stoves remained basically similar between the wars.

Control was by four position switches, on/off and three heat settings. This was achieved by having two elements. Both switched on (in parallel) gave full power, one element gave half power and the two in series gave one quarter power. This wasn't very controllable compared with a gas cooker, so it was at some disadvantage. To overcome this, the electricity undertakings had various schemes. Most of them hired out the stoves, so that the user was not faced with the relatively high cost. This unfortunately had the effect of stifling innovation as they didn't want to be left with older models when new ones became available.

This meant that important improvements like thermostatic oven controls didn't appear until the late 1930s and the simmerstat continuous control for the boiling rings, though invented before the war, didn't really come into use until after it. Despite these drawbacks there wre nearly one and a half million electric stoves in use in Britain by 1939.[16] This compares unfavorably with the eight and a half million gas stoves at the same point.[17] In America, in the following year there were about 2.4 million, which was a lower percentage of homes.[18]

As the electric boiling rings were not very effective at heating water, it was normal to provide an electric kettle with the cooker. There were numerous other devices which sold in smaller numbers, such as grillers, plate warmers, small egg boilers and so on. These were mostly a hangover from the separate breakfast room, but some customers still liked them. Something else that has now disappeared were the electric saucepans.

More important was the electric toaster. This began as a simple device with heating elements that toasted one side of the bread which then had to be turned over manually to do the other side. The user had to watch it carefully to make sure the toast didn't burn. Though automatic devices were available before the 1930s were out, they were not much used, although they were more popular in America where innovations were more quickly accepted.

After having running cold water on tap, the next greatest improvement in life style is to have running hot water. A small minority of houses had solid fuel boilers of one sort or another which could heat a tank using what was known as a "gravity" system. A pipe sloped up from the fire-back boiler to the storage tank. A second pipe brought the cooled water back down. The system depended on the idea that hot water rises and cool sinks. In practice, it required large diameter pipes to work well, and even then required frequent attention to keeping the fire running and ensuring that it didn't overheat and boil the water.

Electricity provided the possibility of a simpler system—just put an electric heater into the tank. With early "immersion" heaters it was necessary to keep an eye on them and turn them off when the water was hot enough, but soon came the idea of adding a thermostat so that the process was automatic and could be left unattended.

In many of the older houses it was too great a task to add a complete hot water system, and adding a local heater over the bath or sink was a popular solution. Many of these were gas geysers which heated the water as it was required without storing it. Often the supply was little more than a trickle if a decent temperature was required, and the problems of lighting them were legendary. (Turn gas on, strike match and insert into hole. Retreat to a safe distance for the ensuing explosion.) Though there were attempts to make electric instant water heaters, the amount of power required would dim the lights for miles around, and the idea had to be abandoned as impractical.

The solution was to have storage water heaters in various sizes: small ones in the kitchen and larger ones in the bathroom. These were quite popular in the late 1930s and more were installed than immersion heaters. Once the Grid was in place by 1933 the numbers of electric water heaters expanded much more rapidly in Britain than in other European countries, from around 30,000 to about half a million at the beginning of the war, though this still was only 4% of households.[19]

So far all these devices depended on electric heating elements, but electricity could do much more, and there were other devices that depended on motors. In 1901, Hubert Cecil Booth, a British engineer, patented his invention of a vacuum cleaner to suck dust and dirt out of carpets and household fabrics.[20] The pumping part was so large that it normally remained outside in the street and the cleaning was achieved with long flexible hoses that were fed in through the windows. In the next few years it became quite a fashion to invite friends to a party while the vacuuming was being undertaken by Booth's uniformed men. It was a great improvement over duster and dustpans and brushes which mostly redistributed the dirt rather than removing it.

Though numerous manual devices appeared roughly based on the same principle, it was an asthmatic janitor in Ohio, America, James Murray Spangler, who made the next step.[21] It was a crude device made from a broom handle, a pillow case and a tin soap box but, crucially, it had an electric motor to drive the fan which created the suction. It was also small and light enough to be used by one person. He might have used it just for his own convenience if his cousin hadn't been married to William H. Hoover (Fig. 12.5).

Hoover soon had an improved device in production and it became a hit in America. By 1912, Hoover were exporting cleaners to other countries, particularly Britain, and began manufacturing there after the war. The rotating brushes were added, powered by the motor which led to their slogan "it beats as it sweeps as it cleans". This laid the foundation for all the "bag on a stick" cleaners that were to dominate the market for over half a century.

The only other forms that appeared were the "cylinder" cleaner pioneered by the Swedish company Electrolux, and a curious spherical device from Booth's British Vacuum Cleaner company which was sold under their Goblin brand name. These devices depended on pure suction, rather than having brushes, to help raise the dirt from the carpet.

Vacuum cleaners were ideal items to be sold door-to-door. They were light enough for the salesman to carry and easy to demonstrate once the electrification of houses reached a satisfactory level. Though undoubtedly a useful device, the rapid take-up was accelerated

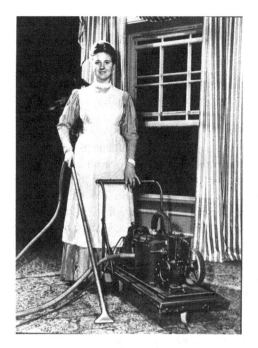

Fig. 12.5 (*Left*) An early "portable" vacuum cleaner. The maid rightly looks terrified of it. (*Right*) An early Hoover derived from the Spangler design. It is a much more consumer-friendly device. Source: Byers A, *The Willing Servants*.

by the cut-throat competition between the teams of commission-hungry salesmen. Their success was shown by the numbers of homes that had one, which even in the UK—well behind America as usual—had reached about 2.3 million by 1939.[22]

There the designs stuck, and were little altered until James Dyson introduced the cyclone vacuum cleaner in the 1990s. The industry had become completely fossilized and made a good living by selling replacement bags to collect the dirt in their machines. It was a cushy number and they didn't want it disturbed by Dyson's revolutionary bagless design. In the end their opposition proved a mistake and his machines swept the market.

The opportunities offered by electric motors were exploited in other devices. The most obvious was in tackling the drudgery of washing. Washing machines appeared early in the century, and though popular to some extent in America, made little impression in Britain until after World War II. The same is true of refrigerators. Because of the delayed take-up these will be dealt with in later chapters.

A curious market was created by the standardization of frequency: clocks. An AC synchronous motor runs at a precise speed depending on the frequency of the supply. With suitable gearing, a clock could be made with second, minute and hour hands. Because the frequency on a network such as the Grid needed to be tightly controlled, and was made correct every 24 h, these clocks kept better time than most mechanical clocks and watches. In the late 1930s they became quite popular.

The electricity supply companies didn't leave the promotion of electrical appliances in the hands of the makers. They had a number of their own schemes to try to advance their use. A popular one was to supply the items "free" with the cost being recovered "with an addition to the quarterly electricity account".[23] Also common were "installment plans" or "hire purchase". Here the customer received the item but had to pay for it over so many weeks or months. The problem was that the customer didn't own the device until all the payments had been made and any arrears were used as an excuse to repossess the item. The system was open to much abuse until legislation controlled the worst excesses.

Did all this activity have any effect? The curves above show the dramatic increase in the number of houses supplied with electricity, but for the industry this wasn't enough. The domestic customers were expensive to service, so they were looking to increase their usage. As Fig. 12.6 shows, the average usage per household in Britain connected with electricity didn't change much from the end of World War I until 1933. In this time most of the power was used for lighting. Though some of the early, and hence richer, customers were starting to use more electricity as they obtained appliances, this was counterbalanced by new customers who probably had smaller establishments and used less.

Once the Grid was in place the average usage started to rise, and continued to do so steadily until the next war. In the UK, all the factors had come together and a virtuous circle created. The Grid had driven down the cost, and this encouraged more users and also the increased use of appliances.

In America, the pattern was slightly different as the advent of electricity in the home got going rather earlier and the take-up of appliances followed that, but the impact of the

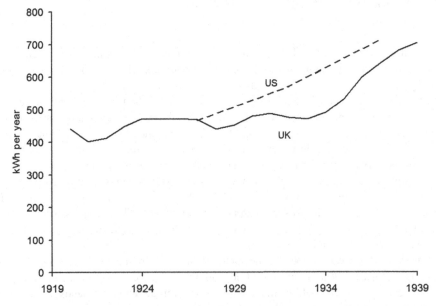

Fig. 12.6 Electricity consumed by the average American and British connected household, 1919–1939, showing remarkably similar patterns. In the UK, the rise began after the Grid was finished. Source: Author.[24]

Depression in the 1930s was much more severe, stalling progress. The result was that, by the late 1930s other countries, particularly Britain, were not so far behind.

However, the consumption per household in the two countries was remarkably similar. Though there is a shortage of the data, up to 1927 they are likely to show much the same lack of change. After that date the US case started to increase whereas the UK didn't begin for some years. It is noticeable that in America the consumption per household was increasing despite the Depression. For those in work, and they were likely to be the ones with the electric supply, it was not such a bad time and they started to get electric irons, vacuum cleaners and so on, hence using more electricity.

In this interwar period, which is often regarded as one dominated by the Depression, electricity drastically altered the way of life of so many people. In that time some fifty-five to sixty percent of an increasing number of homes had been wired for electricity. Much of the drudgery that women suffered in the home had been or could be reduced. Despite this, the time spent had not reduced as much as might be expected because standards improved. Electricity in the home had come of age and had a bedrock on which to build once the war was out of the way.

NOTES

1. Butchart E. Unemployment and non-employment in interwar Britain. University of Oxford, *Discussion Papers in Economic and Social History*, 16, May 1997.
2. Matthews R.B. Industrial, agricultural, and domestic applications of electricity, including illumination and tariffs. *Journal of the IEE*, 72:434, 132, 1933.
3. Scott P. Consumption, consumer credit and the diffusion of consumer durables. In F. Carnevali & J-M. Strange (eds), *20th Century Britain*, p. 171.
4. Houses built from B.R. Mitchell, British Historical Statistics, p. 392; houses wired for electricity by subtracting the annual values in L. Hannah, *Electricity Before Nationalisation*, p. 188 with some adjustments where the swings were too extreme.
5. All these curves should be approached with caution as there are no reliable figures and they are based on estimates of varying degrees of reliability. The sources are: Houses wired from Hannah, p. 188 with small adjustments; cookers calculated from: 1923 Hannah, p. 193; 1927/8 Morley C.G. Domestic applications of electricity, *JIEE*, 68, 147 1930; 1931 Pugh M. *We Danced all Night*, p. 180 (this has slipped a decimal point); 1932 Bowden S.M. The consumer durables revolution in England 1932–1938, *Explorations in Economic History*, 25, 42–59, 1988; 1934–39 PEP, Report on the market for household appliances, p. 69; vacuum cleaners calculated and extrapolated from annual production figures in PEP, p. 211; electric irons calculated and extrapolated from: Bowden S. & Offer A. Household appliances and the use of time. *Economic History Review*, XLVII, 4, 725–748, 1994.
6. Bellis M. History of the oven from cast iron to electric, available at: http://inventors.about.com/od/ofamousinventions/a/oven.htm.
7. Snodgrass E.M. *Encyclopedia of Kitchen History*, p. 343.
8. *The Times*, Saturday, August 6, 1892; e.g. *Hull Daily Mail*, Monday January 16, 1893.
9. US patent 811,859.
10. DNB, Belling, Charles Reginald.
11. See the censuses or, for a summary, Woollard M. The classification of domestic servants in England and Wales, 1851–1951. Proceedings of the Servant Project, vol. II, Seminar 2. Oslo, 13–15 June, 2002.

12. *Dover Express*, 15 February, 1924.
13. Data from: American homes wired for electricity 1920–1956, available at: www.mnhs.org/school/historyday/lessons/images/tabletext2.pdf; Bowden S. & Offer A. Household appliances and the use of time. *Economic History Review*, XLVII, 4, 725–748, 1994; Tobey R.C. *Technology as Freedom: The New Deal and the Electrical Modernization of the American Home*, p. 7.
14. PEP, Report on The Market for Household Appliances, p.125 quotes a survey by the Women's Advisory Housing Council which suggests that nearly 55% of homes had an electric fire in 1939.
15. GB patent 108,827 1917.
16. PEP, p. 69.
17. PEP p. 67.
18. Tobey, p. 7, times the number of households.
19. Bernard J.I. The application of electric heating to domestic hot-water supply systems. Journal IEE, 85:511, 1939.
20. GB patent no 1901 17,433.
21. US patent 935,558.
22. Scott P. Managing door-to-door sales of vacuum cleaners in interwar Britain. *Business History Review*, 82, 761–788 2008.
23. E.g. The Cheltenham Electricity Supply, *The Times*, January 13, 1934.
24. Calculated from: UK: Domestic electricity from Dept Energy and Climate Change, 2332 historical electricity data 1920–2010, (the figures are in close agreement with Hannah, p. 428); Wired houses from Hannah, p. 188 as above; US: Quinquenial US special censuses of the electrical light and power industry 1927–1937, available at: http://eh.net/database/quinquennial-u-s-special-census-of-the-electrical-light-and-power-industry-1902-1937/; wired houses from American homes wired for electricity 1920–1956, available at: www.mnhs.org/school/historyday/lessons/images/tabletext2.pdf.

13

Blackout: War and Crisis in Electric Power Generation

You can't fight a war and scrape right down to the bottom of the barrel, throwing in everything you've got, and then start up again as if nothing had happened.

Clement Attlee

After World War II, electricity supply was generally taken for granted. There were, however, a couple of cases that gave considerable cause for thought. In Britain it was the overhang of war that led to disaster, while in the US it was a fragile network that had grown organically that proved vulnerable. These two events together show the position electricity had reached as an essential part of life which developed nations could not function without.

One Friday afternoon, nearly three hours into a British House of Commons debate about the shortages of coal, the Minister of Fuel and Power stunned the Members with a rambling statement which drew low whistles. The essence of the announcement was that power stations in the London, South-Eastern, Midland and North-Western districts had insufficient coal to keep supplying the current demand. Subsequently, as from Monday, no electricity would be supplied to industrial consumers in those areas, and supplies to domestic consumers would be cut off from 9 a.m. to 12 noon and from 2 p.m. to 4 p.m. in the afternoon.[1] It was the lowest point for the electricity industry since it began.

The date of this was not, as might be expected, in the middle of the war; it was February 7, 1947. The hapless Minister reluctantly making the announcement was Emanuel Shinwell, who was under pressure for not acting long before this. He seemed to think these restrictions would not be needed for long, but in the event the industrial ones lasted for three weeks and the domestic ones for seven. In practice, essential services were exempt and many households had to be put on their honor as it was impossible to switch them off without affecting hospitals and other essential services.

It had started snowing on January 23, and though it didn't come down every day it lay on the ground in many parts of the country for the next two months with temperatures rarely rising above freezing even in the daytime.[2] At midnight on January 28, Big Ben struck once and then stopped. The next day, part of the Thames froze.[3] In was the coldest winter for more than 50 years.

What finally precipitated the crisis was that 57 ships loaded with coal for the south were stuck in ports in the North-East by gales, and in South Wales the docks were frozen. The railways were seriously disrupted because of the snow, and the vital Woodhead tunnel linking the North-West with Sheffield was blocked. Miners were unable to get to their work in the collieries due to the weather conditions.[4]

For ordinary people the conditions were terrible and some felt that it was worse than wartime. Then there had been the collective spirit and the sense of fighting a common enemy, now there was none of that; it was just hell. The country was exhausted, and this fresh blow was the final straw for many.[5] Out came the candles and lamps that many thought they had put away for good, and the shock of turning off the street lights to save power brought back all the old wartime fears (Fig. 13.1).

With few coal deliveries, very low pressure for the gas, and the electricity off for long periods, the problem was to stay warm. In South Wales James Lees-Milne grumbled, "All my pipes, including wc pipes, are frozen, so a bath or wash is out of the question. Wc at office frozen likewise… And we live in the twentieth century. Even the basic elements of civilization are denied to us."[6]

In London, Maggie Joy Blunt wrote in her diary, "Candles and lamps in shops. Candles in café where I went for coffee and candles and oil lamps in local shops too—in local library only central room is heated, side rooms are icy… Wartime gloom on every face. Scarves, mufflers, wool and fur gloves, fur-lined boots and bootees, fur coats, all much in evidence… A feeling of 1930s panic prevails."[7]

Factories soon began to close down through lack of coal or electricity, throwing many people out of work. The huge Austin car works in Birmingham shut down, while in Lancashire the effect on the cotton industry was severe. This was a double disaster as the

Fig. 13.1 Italian prisoners trying to clear a road of snow in Leicestershire in the winter of 1947. Source: Author.

country had a severe balance of payments crisis and needed every bit of export that could be achieved. By February 22, more than two million people were out of work.[8] The figure was to peak at 2.37 million before dropping as the weather conditions improved.[9]

The seeds of this disaster went back some seven or so years to World War II. Unlike World War I, conscription did take place early on but it was not coupled with a manpower policy. The result was that large numbers of skilled people were stripped out of the electrical and mining industries. This didn't seem to matter at the time, but gradually the effects became more and more serious. On top of that, resources of materials were totally committed to the war effort which, considering that it took four years to build a power station, was storing up trouble for the future.

For the meantime, the watchwords were "make do and mend", and this applied to the electricity supply as well as everything else. As the production of munitions accelerated, so the need for electricity increased. To keep away from the vulnerable eastern areas the bulk of the factories were cited in the West Midlands, south-west England and South Wales, where there wasn't enough electricity generating plant (Fig. 13.2).

The combination of evacuations, blackout, and the run down of some peacetime production meant that the electricity demand in London and the southeast declined sharply and wasn't to recover to its 1938 level until the end of the war.[10] The obvious thing was to ship the power westwards on the Grid, but it had been designed as an interconnection system and not for wholesale power transfers.

The government was forced to agree to extensions and reinforcements to the Grid to achieve this, together with a small amount of additional plant in power stations under construction. The problem was that the industry had lost a lot of its skilled manpower and the

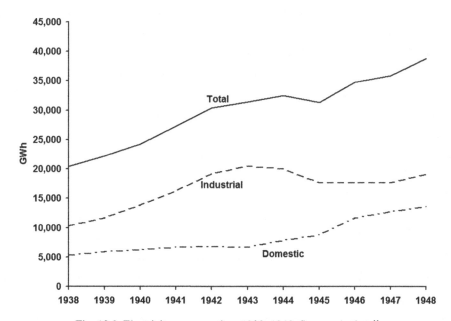

Fig. 13.2 Electricity consumption, 1938–1948. Source: Author.[11]

Grid improvements in 1940 and 1941 were completed by "a motley crew of recruits from outside the industry such as clerks and waiters".[12]

One of the things that saved the industry was that the nature of the load changed. The lighting blackout, brought in to make it more difficult for enemy bombers to find their targets, had the effect of reducing the evening peak in demand. As factories geared up for war shift, and in some cases continuous, working spread the time of day when the power was taken. While the amount of electricity for industrial use rose markedly, that for domestic consumption largely leveled out. The peak demand, the critical factor because it defined what the generation system had to supply, didn't increase anything like as fast as the number of units used (Fig. 13.3). The Load Factor improved and so the system was largely able to cope.

Before the war the Central Electricity Board (CEB), in charge of the Grid, had made plans to deal with the impending conflict. The supposition was that in the event of power stations being damaged, the Grid could continue to supply customers as long as it was kept working. Thus the CEB built up a stockpile of transformers, switchgear, temporary pylons and cable and this enabled the disruptions to the system to be rapidly repaired.

Damage by enemy action was surprisingly light, and most was repaired in a day or so. At no time was the loss of capacity greater than about four percent. The most serious event was when bombs fell on Fulham power station on September 9, 1940 which turned the plant into a terrifying mass of escaping steam and put it completely out of service. The Grid connection was undamaged and so it was able to go on supplying customers. A third of the station's output was restored after nine weeks, two thirds after 15, and the whole output after 28 weeks.[13]

The Grid itself suffered from a few hits, notably in Plymouth where a substation was destroyed, and occasional damage to power lines. The spares from the stockpile proved

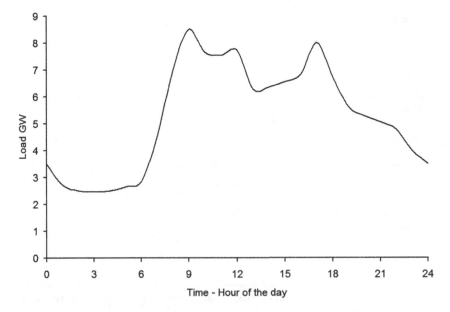

Fig. 13.3 Typical daily demand curve. During the war the evening peak reduced and the morning one risen as demand changed from lighting to industrial use. Source: Author.[14]

their worth but the problems from enemy action were only eight percent of wartime faults, considerably less than had been anticipated. The biggest difficulty was from what was categorized as "defensive action". This consisted of damage caused by a number of things: friendly aircraft hitting power lines, shrapnel from anti-aircraft guns, and military exercises in which overhead lines were damaged by rifle, machine-gun, tank-gun and mortar fire, flamethrowers and even hand grenades.[15]

The biggest threat was from barrage balloons; their trailing steel cables breaking free accounted for thirty-five percent of all faults on the system.[16] Though they might have been valuable as a defense against low-flying aircraft, they were a menace to the Grid. Despite this, the number of interruptions to the supply was quite limited. There were three in the winter of 1940–1941, nine the following year, and six in the winter of 1944–1945. Usually they only affected two or three of the Grid's districts and often only for 15 min or so, although they could be as long as 3½ h. Only once were cuts necessary in all the Grid's seven districts.[17]

In 1943, the industrial load hit a peak largely determined by the demand for munitions and other war-related production. This declined slightly in 1944 and then more sharply in 1945 as the manufacture of these items scaled back. Even though the numbers employed were still largely the same, the reduction in the supreme effort that had characterized the early years of the war is clear to see in the reduced electricity consumption.

Domestic demand went the other way. In 1944, after the long period in the doldrums, it began to rise. The major cause of this was the shortage of coal brought about by a lack of miners because they had gone to war, and that the available supplies were required for the war effort. Coal for burning in the grate at home was the lowest priority and so very difficult to obtain.

Fortunately for the electricity supply industry the decline in industrial demand in 1945 was greater than the rise in domestic consumption. This led to the first drop in electricity output since the post-First World War slump in 1921. Thus by luck, and a good deal of juggling, the system had got through the war, and fulfilled what was required of it, in a rather better fashion than had been feared at the outset.

However, there had been very little construction of generating plant over the war years, and when obsolete plant was closed down this made the situation worse. Though there is some confusion about the exact amount of increase, what is clear is that with the reduced wartime building there wasn't going to be enough capacity in peacetime.[18] The CEB assumed that a brief post-war boom would be followed by a slump as had happened after the previous war, but they were still worried that they wouldn't be able to meet the rising demand.[19] At some point their luck was going to run out.

From the end of the war until late 1946 around a quarter of a million extra houses were in use. This included large numbers that had been so badly damaged that they were uninhabitable during the war being brought back into use, as well as new temporary (prefabs) and permanent buildings. Most of these had an electricity supply, hence producing extra demand. From the beginning of 1945 existing homes were again being connected to the supply, making a total of 400,000 new consumers each year.[20]

Demobilization went remarkably smoothly with industry and commerce absorbing the manpower as it became available. Unemployment hardly rose. Women, on the other hand, often went back into the home. While there had been 7.2 million women in work at the peak in wartime, this had fallen to 5.8 million by September 1946.[21] More people in the home meant more home fires to be kept burning, even if they were only electric ones.

With the soldiers paid off and with civilian jobs available it was a time to spend on whatever was available. So many basic items, particularly food, were still rationed it was natural to buy whatever was obtainable. The result was that the cinemas were full, reaching their highest audiences ever.[22] The dancehalls were throbbing, but many people were looking to make the life at home easier. One thing that was not on the ration, or controlled, was an electrical appliance. The manufacturers saw their chance and the buyers responded.

It began in 1945, but got into full swing in 1946. Electric cookers and wash boilers sales were modest, in the order of 200,000, water heaters around 450,000, electric kettles and vacuum cleaners well above 600,000. More startling were the number of electric irons made at around 3.4 million, but most damaging to the electricity supply were the nearly three million electric fires.[23]

A one-bar electric fire has a rating of 1 KW, but many electric fires had more than one bar. If everyone turned on just one bar then it added 3000 MW to the load on the Grid which was capable of supplying a total of around 8000 MW. The possible load in the country is always much larger than the capability of the supply. The system depends on not everyone turning everything on at the same time.

One of the factors driving the rush for electrical appliances was that electricity was relatively cheap. The cost of coal had been rising steadily, but to try to restrain potential inflation the government had held the price of electricity. Many in the industry thought that the price had been too low in any case because it reflected historical costs and not what the industry would need to face. The result of this underpricing was that it was as cheap to heat with an electric fire as it was to burn coal in a grate.[24]

The CEB survived 1945 reasonably well, but there was a huge backlog of maintenance from the wartime. By the autumn of 1946, much of the plant needed refurbishing, and the signs were very worrying (Fig. 13.4). Demand was rising fast and the peaks were going beyond the available capacity leading to what was euphemistically called "load shedding", which in common parlance means "power cuts". As Christmas approached these were becoming more and more common.

The available capacity was a long way below the rated output of the system, due to plant breakdowns and all sorts of miscellaneous difficulties which meant that the full output couldn't be obtained. Perhaps the most frustrating problem was the poor quality of the coal. This had been a cause of worry throughout the war. There was a story that an incendiary bomb had fallen on a coal stock pile, but far from it bursting into flames, it put out the device.[25] In late 1946, the problem was still continuing, and some hundreds of MW of potential output were being lost because of the poor fuel.[26]

The CEB was also concerned at the low stocks of coal. Before the war they had normally aimed to enter the winter with 10–12 weeks of stock, but now this was totally impossible and they thought the government's target of six weeks supply disastrously low. In 1946 they entered the winter without even 4 weeks' supply.[27] It was a dangerous gamble.

The scene was now set and all it needed was a freeze. The weather obliged. As December 1946 wore on, load was being shed daily on the peaks, reaching as much as ten percent of the available output.[28] At the turn of the year, the weather was a little milder but the system still couldn't cope with the peaks. When it began to snow later in January, the game was up.

Every day on the peaks greater and greater amounts had to be shed as the customers turned on all those electric fires to keep warm. Finally, on January 31 the demand hit an

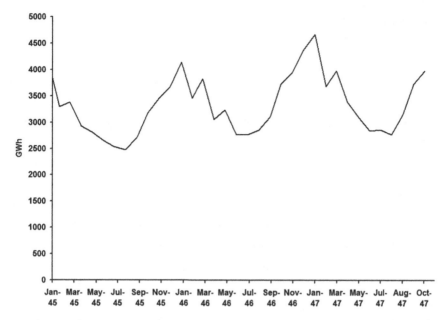

Fig. 13.4 Electricity supplied each month showing the sharp peaks in December 1946 and January 1947 before the restrictions came into force. Source: Author.[29]

estimated peak of over 11,000 MW, but with only about 9000 MW of capacity available. Even if absolutely everything had been in service, this load couldn't have been met.[30] Extensive and long duration cuts were inevitable (Fig. 13.5).

As if this wasn't a severe enough problem, by early February the coal stocks had fallen to an average of 1.5 weeks, and in some places they were virtually exhausted.[31] The freeze, outlined at the beginning of the chapter, took hold and the coal that had been mined was unable to be moved in many areas. The CEB was facing total disaster with the potential complete collapse of the Grid, plunging the whole country into darkness.

The CEB went to the Minister of Fuel and Power proposing a drastic curtailment of use. On February 5, the Minister presented a paper to the Cabinet outlining all the problems reported to him, but ended with the recommendation that no action should be taken yet![32] Two days later, after taking a hammering in the House of Commons, he introduced just the measures the CEB had recommended.

The reductions soon eased the pressure. The consumption of coal, which averaged about 500,000 tons a week throughout 1946, had climbed to over 700,000 tons a week in late January and early February. Once the measures came into force it dropped back to 500,000 tons a week, even though they didn't apply to the whole country.[33]

The demand curve throughout the day now took on a very strange form with peaks in the early morning, and at midday and evening, with rapid reductions in between. This gave the CEB considerable control problems, particularly in London which had a high domestic load. The drop from the high morning peak was very sharp in the South East as the measures took effect, and only by allowing the frequency to swing to the limits of the acceptable range, and importing and exporting large amounts of power over the Grid at the appropriate moments, were they able to keep the situation under control.[34]

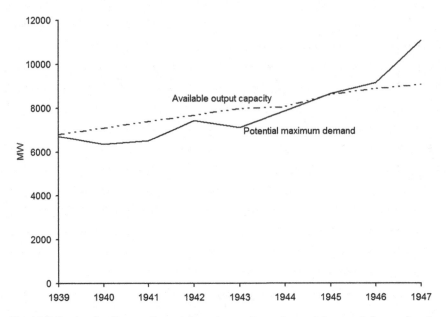

Fig. 13.5 Recipe for disaster. Potential maximum demand was rising much faster after the war than the limited available capacity. The dates refer to the winters of each year, as the peak usually fell in January. Source: Author.[35]

Fortunately, even the morning peak demand was much reduced and usually came within the system's capability or could do with small amounts of load shedding. Gradually the coal stocks increased and, with a restriction on the use of electric space heating remaining in force over the summer, the situation was got under control. Though there was a considerable amount of load shedding in the autumn of 1947 it was manageable and the relatively mild winter meant that it passed off without too much pain.

All this had given the electricity supply industry a nasty shock. It was now determined to build sufficient plant not to get into this position again, but faced government control of resources and the four-year cycle time needed to bring new power stations into service. It was going to take a while.

It was hoped that the nationalization of the mines at the beginning of 1947 would gradually solve the coal supply problem, but in the longer term an alternative source of power was needed. As for the consumers, it didn't put them off. They put it all down to the overhang of the war, and as the cuts had been spread about nobody had suffered too much. Electricity had become too much part of daily life for this to dissuade them. Most people didn't know how close the system had come to disaster.

The US, however, had a stroke of luck when it came to the war. Those great hydroelectric schemes, including the Hoover and Grand Coulee dams and the Tennessee Valley Authority, had all come on stream. They had originally been undertaken to provide employment in the depths of the Depression, but were now a national asset. By 1940 some 1500 hydroelectric schemes provided about a third of the country's electrical energy.[36]

Even so, with the ambitious plans for war production it soon became apparent that the amount of power available would be insufficient. It was claimed that the Axis powers had three times the amount of power available. Fortunately, the generation capacity at many of these dams could fairly easily be expanded, leading to a considerable increase in power plant during the war. The energy-intensive production of aluminum, so vital to aircraft manufacture, could be expanded to meet the war requirements.

With no fear of bombing or threat of invasion, the west coast became a haven for ship-yards, steel mills, chemical companies, oil refineries, and automotive and aircraft facto-ries, all needing vast amounts of electrical power which could easily be supplied, particularly from the Grand Coulee installations. The siting of the Hanford and Oak Ridge plants of the Manhattan Project, for producing atomic weapons, was no accident. They were near these sources of low-cost electricity.

The power from the hydroelectric schemes was sold to the electricity utilities to distrib-ute. Extra high voltage overhead lines were built and the various power supply networks grew organically. The electricity industry, like so many others, benefited greatly from the war. After it, with the vast majority of homes, including a high proportion of farms, now with a supply, things looked to be set fair.

In each of the two decades after the war electricity consumption roughly doubled.[37] Prosperity turned the American home into a palace of electricity consumption with appliances everywhere. Ever-increasing numbers had washers and dryers, televisions, coffee makers, dishwashers and, particularly, air conditioning. It was driven by cam-paigns such as General Electric's "Live Better Electrically", which encouraged the all-electric home.

Offices and factories had also been transformed. They now depended totally on elec-tricity for lighting, air conditioning and running all the working equipment, be it typewrit-ers and adding machines or production tools. Tall buildings were only possible because of electric elevators. Manufacturing was only practical with abundant sources of electricity. Shops and stores needed lighting, cooling and escalators to function. In fact, particularly in the cities, everyday life was utterly dependent on it.

To meet all this increased demand, the power industry in the north-eastern United States had interconnected their systems to produce "grids". There were two in the area, the Ontario-New York-New England pool (The Canada-United States Eastern Interconnection, or CANUSE area), which drew some of its power from the Niagara Falls, and the Pennsylvania-New Jersey-Maryland pool (the PJM interconnection). Together these made up the north-east power grid.

At dusk on November 9, 1965, eleven-year-old Jay Hounsell was making his way home in Conway, New Hampshire.[38] As young boys do, he was hitting the street light poles with a stick as he passed when one of them went out. Then he realized that it didn't stop there, the whole town was now dark. He ran home in panic, thinking that his whack with the stick had been the trigger of the whole thing.

In reality, the cause was just a protection device on 230 kV overhead power line near Ontario, miles away in Canada, that had not been set correctly. This created a power surge which quickly overloaded lines in New York State and the fault began to "run". First Buffalo, Rochester and Albany went and within a few minutes the blackout spread to Boston, Connecticut, Vermont, part of Canada and then Manhattan, the Bronx, Queens and most of Brooklyn. When it stopped, some 30 million people in eight US states and the Canadian provinces of Ontario and Quebec were blacked out.[39]

The timing was perfect for maximum disruption as it was at the height of the evening rush hour. Millions of commuters were delayed and some 800,000 people were trapped in New York's subways, with thousands more stranded in office buildings, elevators and trains. For many, the long walk home was not practical and they were forced to try to find somewhere to shelter. The hotels quickly ran out of rooms and people bivouacked in their lobbies. Some 10,000 would-be travelers were stuck in subway cars, unable to escape the dark tunnels and by midnight the Transit Authority was sending food and coffee down to them.

On the surface, 10,000 National Guardsmen and 5000 off-duty police were called in because looting was expected. Neither this nor rioting broke out; it was remarkably peaceful. Volunteers helped direct traffic as the traffic lights no longer worked, and assisted the Fire Department to rescue stranded people. Others shared candles and flashlights. For those who thought they could go by road there was a shortage of fuel as the gas stations' pumps didn't work.

It wasn't a simple matter to get the power restored, and by 11 p.m. in only a few areas had it come back on. It took until 2 a.m. before the whole of Brooklyn was supplied again. But it was just before 7 a.m., nearly fourteen hours after the failure, before the whole of New York was powered up again. One of the difficulties was that, without power, some of the generating stations, which had tripped out because their load had disappeared, couldn't be started up and portable generators had to be brought in to get them going again.

It was a complete shock to everyone. President Lyndon Johnson said it was "a dramatic reminder of the importance of the uninterrupted flow of power to the health, safety and well being of our citizens".[40] The electricity supply industry suddenly realized that it had not taken a sufficient interest in the reliability of the system. Though the immediate cause was soon identified, is was also found that many of the lines were too heavily loaded, and many other similar problems came to light. It was time for a considerable rethink.

Other countries, like Britain, had some problems with running faults but nothing remotely approaching this scale. They usually had centrally planned and controlled networks and in many cases a single, state-owned organization which didn't depend on numerous utility companies having to cooperate, which they didn't always do to the necessary level.

NOTES

1. Hansard HC Deb 07 February 1947 vol 432 cc2122–206.
2. Weather data from: G. Booth, Winter 1947 in the British Isles, available at: http://www.winter1947.co.uk/.
3. Pimlott and Dalton, p. 478.
4. Cabinet Papers, Feb 5, 1947, CP (47) 50.
5. Kynaston D., *A World to Build*, p. 192.
6. Quoted in Kynaston, p. 190.
7. Garfield S., *Our Hidden Lives*, p. 352.
8. Hansard, HC Deb, February 24, 1947 vol 433 cc1704–10.
9. Hansard, HC Deb, March 10, 1947 vol 434 cc963–1094.

10. Hacking J. and Peattie J.D., The British Grid system in wartime. *JIEE*, 94: 41, 463–476, 1947.
11. Data from Hannah, *Electricity before Nationalisation*, p. 428 which are based on Ministry of Fuel and Power Statistical Digest 1948 and 1949. However, these don't quite agree with the figures in the Appendices of the First Report of the British Electricity Authority 1949. The discrepancy would appear to be due to the generation and supply of small systems outside the control of the BEA.
12. Hannah L., *Electricity before Nationalisation*, p. 296.
13. Hannah, p. 293.
14. Data based on A.R. Cooper, Load dispatching and the reasons for it, with special reference to the British grid system. *JIEE* Part II: Power Engineering, 95:48, 713–723, 1948.
15. Hacking and Peattie.
16. Hacking and Peattie.
17. Hacking and Peattie.
18. Hannah on p. 386 says 3896 MW; in his capacity tables on p. 433 the increase calculates as 2812. Cooper's figures give 1696, and Hacking and Peattie gives 3091. It very much depends on the exact start point as plant was being installed up to the start of the war.
19. Hannah, p. 307.
20. Hannah, p. 324.
21. Kynaston, p. 99.
22. See Chapter 7.
23. Central Statistical Office, Monthly Digest of Statistics, December 1947, Tables 69 and 86.
24. Hannah, pp. 325–326.
25. Robertson, J.N., Discussion on the British Grid system in wartime. *PIEE*, 96: Part II, 25.
26. Cooper.
27. Hannah, p. 315.
28. The pattern of load shedding is from Ministry of Fuel and Power 1947–48 [Cmd. 7464] Report of the committee to study the electricity peak load problem in relation to non-industrial consumers.
29. Data from Central Statistical Office, Monthly Digest of Statistics, January 1946, Table 33 and December 1947, Table 32.
30. Central Electricity Board Annual Report 1947, p. 45.
31. Hannah, p. 316.
32. Cabinet Papers, February 5, 1947, CP (47) 50.
33. Average computed from Hannah, p. 433, January from Hansard *HC Deb, February 7, 1947, vol 432 c2171*, others from Hansard *HC Deb, March 5, 1947, vol 434 c491*.
34. Cooper.
35. The data for this graph have been compiled from three sources: Hacking and Peattie, Cooper, and Central Electricity Board Annual Report 1947, p. 45.
36. Bureau of Reclamation, The History of Hydropower Development in the United States, available at: http://www.usbr.gov/power/edu/history.html.
37. Great Northeast Blackout, available at: http://www.blackout.gmu.edu/events/tl1965.html.
38. Biever, R.G., 1965 blackout a part of 1960s pop culture, available at: http://www.electricconsumer.org/1965-blackout-a-part-of-1960s-pop-culture/.
39. This Day in History, The Great Northeast Blackout, available at: http://www.history.com/this-day-in-history/the-great-northeast-blackout; Pressman, G., Remembering when the lights went out in 1965, available at: http://www.nbcnewyork.com/news/local/When-the-Lights-Went-Out-in-1965-133488738.html.
40. Vassell, G.S., Northeast blackout of 1965. *IEEE Power Engineering Review*, January 1991.

14

Give Someone a Bell: Telephones

The day is coming when telegraph wires will be laid on to houses just like water or gas—and friends will converse with each other without leaving home.

Alexander Graham Bell
(1876, shortly after the invention of the telephone)

In 1876, the United States celebrated the centenary of its foundation by hosting the International Exhibition of Arts, Manufactures and Products of the Soil and Mine, known as the Centennial Exposition. It was held in Philadelphia and was one of a series of World Fairs where countries showed off to their competitors. It was opened by President Grant, accompanied by Emperor Dom Pedro II of Brazil. After the speeches, raising of the flag, the anthem and the 100-gun salute, the party went through the Main Hall along a passage lined with soldiers, outside and then into the Machinery Hall. Here they ascended the platform in the center and President Grant and Emperor Dom Pedro simultaneously turned the wheels to set the 1400 horsepower Corliss steam engine into motion.[1] The shafting along the sides of the Hall, and all the belts powering the rows of machines, sprang into life.

With all this pomp and spectacle it was unsurprising that a 29-year-old amateur inventor, not even in the Machinery Hall but in amongst other scientific instrument makers in the Main Hall, should be overlooked.[2] He compounded his difficulties by retreating to his other stall amongst the exhibitors of the Massachusetts Education department in the East Gallery. He probably felt more comfortable here representing Clarke's College for Deaf-Mutes and showing his system of "visible speech" designed to help those with hearing problems, because Alexander Graham Bell was a specialist teacher of the deaf and the professor of vocal physiology and elocution at Boston University.[3]

He had been born in Edinburgh into a family of speech therapists, but following the death of two of his brothers from tuberculosis, the family had emigrated to Canada. From there, Alexander had migrated south to the United States. It was his work, helping people with speech problems, particularly the deaf, which gave him a head start for his invention and its exploitation. Firstly, he understood sound and what began his interest was to try to make a device that would visualize attempts at speech which would help the student to articulate better.

© Springer International Publishing AG 2018
J.B. Williams, *The Electric Century*, Springer Praxis Books, DOI 10.1007/978-3-319-51155-9_14

The second advantage was that the fathers of two of his students, Gardiner G. Hibbard and Thomas Sanders, were influential men who were prepared to back his experiments. They financed him in exchange for a share of the rights to any patents he might produce. They were particularly interested in finding ways to pass more than one message at the same time down a telegraph line.

In 1875, he patented a method based on tuned reeds set to different frequencies.[4] As the reed vibrated it was arranged to make and break an electric circuit. A reed tuned to the same frequency at the receiving end was arranged to indicate the presence of that signal. This will work, but when two or more are put on to the same line they will start to interfere with each other. In an extreme case, the current will flow all the time and it is impossible to separate the signals. At the time, like everyone else, Bell believed that telegraph signals had to be either on or off.

About three months later, while experimenting with this set-up using a number of reeds, Bell was in one room listening to the receiver and his assistant, Thomas Watson, was in another room tuning the transmitting reeds. A reed stopped vibrating so Watson plucked it to start it vibrating. Bell clearly heard a twang. He rushed into the other room to find out what Watson had done. It turned out that the screw was turned down too tightly and the circuit wasn't being broken when the reed vibrated. This was a case a receptive mind appreciating the significance of a lucky accident.[5]

What Bell realized was that, instead of an on or off current being transmitted, a varying one in tune with the frequency of the reed had been produced. This was due to the metal reed altering the magnetic circuit of the electromagnet that was used to vibrate it. At the receiving end the process was reversed. He also knew from his work with sound that this would cause a membrane, like the eardrum, to vibrate. He reasoned that if a membrane was connected to a piece of metal in the electromagnet's magnetic circuit then a current that mimicked the speech would be produced. A similar device, the other way round, would turn it back into sound at the receiving end.

It was one thing understanding the principle, but quite another to make it work. Watson built a model to Bell's instructions but though it transmitted sounds they weren't intelligible. However, Bell was sufficiently encouraged to continue the work and drop the idea of the Harmonic Telegraph (as it had been called). By February 1876, he felt confident enough to file a patent (or Hubbard to do it for him).[6] Controversy surrounds this, as not only was it filed only two hours before a similar application by Elisha Gray, but it appeared to have been altered to include a variable resistance method to act as the microphone at the transmitting end.[7]

At this point Bell still hadn't really made it work. On the March 10, he was trying a different transmitter where the membrane was connected to a wire that moved up and down in a metal cup containing battery acid. As it did so the current flowing through it changed in time with the sound. Watson was in a second room with the receiving apparatus. Bell then said into it, "Mr Watson—come here—I want to speak to you." Watson appeared; he had heard what had been said. They then swapped places and Watson read a passage. Though the words were loud Bell found them muffled, but at the end he clearly heard Watson say, "Do you hear what I say?"

The problem was that a device containing acid wasn't very practical near people's faces. Bell and Watson returned to the electromagnetic device, refining it until it worked,

Fig. 14.1 Bell speaking into his experimental telephone. Source: http://www.officemuseum. com/IMagesWWW/1876_Bell_Speaking_into_Telephone.jpg

and produced a device that under optimum conditions could produce intelligible speech.[8] It was this device that Bell took to the Philadelphia exhibition, only to find that it didn't generate much interest.

Once again it was his teaching of the deaf which was his salvation. One of his visitors was the energetic Dom Pedro concerning the setting-up of a scheme to teach the deaf in Brazil.[9] Typically, he was now taking two weeks to conscientiously investigate all aspects of the Exhibition. As part of his thoroughness he visited the East Gallery and there spotted Bell, who gave him the receiver to put to his ear, while he took the transmitter to the other end of the gallery and recited Hamlet into it. "'My God, it talks," said Dom Pedro (Fig. 14.1).

Dom Pedro's interest ensured that the judges for Group XXV, which included instruments of all sorts, were also given a demonstration. One of them, Englishman Sir William Thomson, later to be Lord Kelvin, wrote the official report in which he said: "This, perhaps, the greatest marvel hitherto achieved by the electric telegraph, has been obtained by appliances of quite a homespun and rudimentary character. With somewhat more advanced plans and more powerful apparatus, we may confidently expect that Mr. Bell will give us the means of making voice and spoken words audible through the electric wire to an ear hundreds of miles distant."[10] Of course, the judges had to try it with one of them at the transmitting end, but it unmistakably worked.

Having achieved some good publicity it was now a matter of improving the device. The troublesome membrane (it tended to stretch in damp air) was replaced, and by October they managed to get it to work over a distance of two miles.[11] Despite this success, Bell and his backers were beginning to realize the difficulties they faced before they would have a working telephone system. Hubbard was losing confidence and he offered the patent rights to Western Union, the giant telegraph company, for $100,000. In one of the greatest mistakes in business history, they turned it down.

Into 1877, things looked better. Other people were trying to patent aspects of telephones; clearly, they thought there was something in this too. Bell was making sufficient progress to form the Bell Telephone Company with Hubbard and Watson. Two days later Bell married Mabel, the deaf daughter of Hubbard, whom he had taught some years before. Almost immediately they left for an extended honeymoon in Europe, and Bell's involvement, apart from some demonstrations in England, greatly diminished. It was left to the others to get the company going.

They were quite successful and by the end of the year they had installed 3000 telephones, but there was a lot of competition, including Western Union who had had a change of mind. In a defining moment, the Bell Telephone Company sued Western Union for patent infringement. It was a courageous thing to do for the little start-up to take on the giant, but the patent held and Western Union was forced to a deal where they withdrew from the telephone business, handing over their telephone lines to the Bell Company in exchange for a slice of twenty percent of the rentals for the duration of the patent.

There were also those just out to make money, trying to find any invention which could challenge Bell's patent. One such was Globe Telephone which had done a deal with an Italian immigrant called Antonio Meucci who had worked for years refining a device that he used to communicate with his invalid wife. Unable to afford the $250 patent fee, he had taken out a caveat in December 1871 for what he called the "Teletrophone".[12] The Bell company's lawyers and experts soon persuaded the court (probably wrongly) that the device mentioned in the badly-worded caveat wasn't really a telephone.

The Bell Company took out some 600 legal actions defending their patents or suing those who had set up telephone systems in contravention of them. Most companies caved in when threatened or actually sued. It was this concerted defense in the courts (presumably initiated by Hubbard) that protected the Bell Company's future and effectively defined Bell's place in history, although in reality he had little to do with the exploitation of his invention.

There was another technical advance at this point. The weakness of Bell's device was that the transmitter didn't have enough power and the signal was rather feeble. A young immigrant called Emile Berliner, working alone, came up with the concept of a microphone where the diaphragm altered the resistance of the contact.[13] By connecting a battery in series with this, quite large changes in signal could be achieved. It was sufficiently interesting for the Bell Company to buy his patent and offer him a job in their laboratories.[14] He was to make an even greater contribution to the record industry later.

A number of other inventors, mainly Francis Blake and Thomas Edison, came up with variants of Berliner's resistive transmitter. In these, the resistance of a solid or granulated piece of carbon was altered by the membrane which vibrated in sympathy with the speaker's voice. The disadvantage of this arrangement was that it needed a battery, but it produced a much greater signal and was much more satisfactory. To protect their position the Bell Company bought Blake's patent, and later gained access to Edison's.[15]

The company went through a number of name changes and permutations but, now in control of the situation and the technology, it grew to control 240,000 lines by 1892. In 1894 the patents ran out and there was an explosion of telephone companies in the US, with the number of telephones reaching a million soon after the start of the century.[16] The problem was that they were independent and unconnected to each other. Even by 1910 only about half were connected to the dominant Bell system. It wasn't until the government got a grip of this in 1913 that real progress was made, but even then it took until the 1940s until they were all connected.[17]

Within 6 months of the early telephones being demonstrated in England, the Bell Company had set up The Telephone Company Ltd. to introduce them to the UK. In 1879 they opened their first service in London with a small 8-line exchange.[18] By the end of the year they had three exchanges and 200 subscribers, so-called because they had to pay a fee of £20 to subscribe to the service.[19] The users were either professionals, e.g., doctors wanting to speak to pharmacists and hospitals, or businesses such as shipbrokers. Initially, these were separate networks and only later were connected together.

The system developed piecemeal with separate companies being set up in various towns and cities. Gradually these all joined together to form the National Telephone Company (NTC). However, the Post Office saw this as a threat to their telegraph operation and so used its powers to limit the telephone companies' operating licences. During the 1890s it took over the long-distance lines which meant that the NTC was unable to build a truly integrated national network.[20]

At first the take-up of telephones was quite rapid, but it was nowhere near as enthusiastic as in America. In 1900, the US had around twice as many telephones per head of population as the UK. By 1922 this had risen to six times.[21] There has been much discussion about the reasons for this, but amongst the factors are the reluctance of business people in the UK to use the instrument (they would get a boy to pass on their message), general conservatism, the high subscription fee rather than call charges and also the "planning blight" of the Post Office license which was due to end in 1911.[22]

In the event, the Post Office made its intentions clear well in advance and it took over the whole of NTC on January 1, 1912 and set about producing a national network. It soon installed the first automatic exchange based on the Strowager system. In all early telephone exchanges, when the caller lifted the receiver they would get through to the exchange where an operator asked which telephone they wished to be connected to, and then make the actual connection.[23] In the automatic system the caller dialed the required number (for this, telephones needed to be assigned numbers) using a rotary dial. Special multiway switches called Uniselectors would click round for each digit, and when all the numbers had been dialed they were connected to the correct line.

An interesting experiment in a different way of using the telephone was the Electrophone system, set up in 1895.[24] For a subscription of $20 (£5) per year the user could be connected to numerous theaters, music halls, the opera and even churches on a Sunday to hear the sermons. The user was supplied with several sets of headphones so that more than one person could listen at the same time. It was never a roaring success, reaching only 600 subscribers in 1908 when it carried performances from 30 theaters and churches. The coming of radio signaled its death knell and it closed in 1926.

It was only after World War I that significant numbers of people began to install telephones in their homes but growth was still slow between the wars (see Fig. 14.2). In the US, the number of telephones reached one for every 10 of the population before World War I; in the UK, this level wasn't reached until well after the second war.[25] The gap was widening.

One of the barriers to growth in a telephone system is the "someone to talk to" problem, often known as the "network effect". There is no point in having a telephone unless people you might wish to communicate with also have one. It tends to be the case that either everyone has a phone, or no one. In America, there was a greater willingness to try it and this led to a faster uptake, only for the Great Depression in the 1930s to knock it back for a decade.

While the telegraph system was essentially useful for government and business and brought enormous advantages in its time, the telephone, while seemingly similar in technology, is very different in use. It is essentially person-to-person communication without

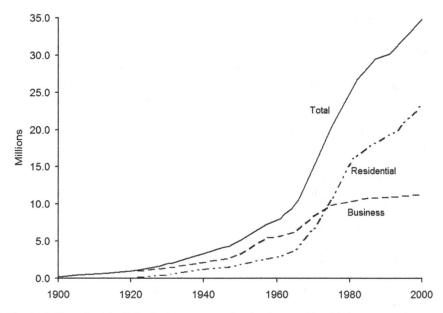

Fig. 14.2 Growth of the UK telephone system showing how residential lines eventually took over from business, 1900–2000. Source: Author.[26]

the intervention of anyone else (apart from the operator in early systems). The telegraph, on the other hand, required skilled operators to generate and read the Morse signals going down the lines. It was only later, to combat the threat to its business from telephones, that automatic machines were used to generate and print the messages.

Thus, while telegraphs were essentially items for offices, telephones were suitable for installation everywhere, including in people's homes. In principle, its potential was much greater. What is surprising is how long it took to fulfill that promise.

One of the problems faced by the early telephone systems was that of working over long distances. With longer and longer wires the signal became weaker and weaker until it could not be heard at the receiving end. With the telegraph it was a simple matter to install a relay which would restore the signal to its former size. There was no equivalent for the telephone until electronic amplifiers became available with the development of vacuum tubes around the time of World War I. Then long-distance calls became possible. In 1915, the first call between New York and San Francisco was made with great ceremony by the now 68-year-old Alexander Graham Bell and his old friend, Thomas Watson.

Around 1920, the number of trunk circuits in the UK (where callers from one local area could talk to those in another) was less than 100, but by 1939 this had rocketed to 6770.[27] The demands of the war meant that shortly after it finished the number had doubled again. For local calls the number of automatic exchanges increased rapidly from a handful in 1922 to over 3000 in 1938, 55% of the total.[28] Despite the initial worries of the Post Office, automatic dialing was popular with the users.

It was now accepted that automation was the way forward, but how this could be accomplished was another matter. It is essential that the number needed to reach someone was the same from everywhere in the country. The problem for a telephone system is that the route needed to get to the end user is different depending on where the call originates.

A partial solution was the "director system" which was introduced in London and some other cities. Each local exchange was allocated a name which was shortened to a three-letter dialing code.[29] The caller, from anywhere in the director area, dialed the code followed by the required number. Each local exchange had a system for determining exactly which route should be followed to get the call to the correct local exchange which would then select the called number. It worked, but the areas covered were limited as the complexity became too great to extend it further.

Gradually over the interwar period the general level of charges decreased.[30] More exchanges were added and the system expanded into fresh areas, naturally bringing in more subscribers. With all the improvements its utility increased, so it is surprising that the growth in this period wasn't greater than it actually was.

It was World War II that made the telephone popular. With an extensive trunk system it was now possible to ring from one part of the country to another by getting the operator to connect you. As people were spread about in the forces, and for war work, the need to keep in touch was much greater and now the telephone system was largely capable of fulfilling this. Understandably, once the war was over there was a greater interest in having a telephone.

In Britain, despite a heavy investment program, the Post Office wasn't really ready for this surge in demand.[31] It installed telephones at around 300,000 a year, double the rate before the war, but it wasn't keeping pace and the waiting list reached half a million in 1951.[32] After that it slowly started to fall, but it took over a decade to clear the backlog.

One of the ways that the Post Office coped with the demand was by getting two subscribers to share a "party line". Each person had a separate telephone and a different number, but shared the same pair of wires back to the exchange. This meant that only one person could use the phone at a time, and that either could hear the other's conversation by picking up their phone. By around 1960 a million people were sharing.[33] Party lines weren't liked, but in these years it was often that or not having a phone at all.

Without these restraints, and with the numbers already high enough to overcome the network effect, American telephone usage in the home really took off. In the following couple of decades, the vast majority of homes became connected. Telephones became the norm. Ownership was approaching one hundred percent well before the advent of the mobile cell phone and the change from phones being tied to a place to being associated just with a person.

In Britain during the 1950s, the number of business telephones increased quite rapidly, but home ownership only continued to grow at much the same rate as before the war. This suggests that, where it had to ration its resources, the Post Office was deliberately giving preference to business users. After about 1960 things began to change. The business demand leveled out while the domestic took off.

It is extraordinary that it had taken almost 90 years from the telephone's introduction into the country for just twenty percent of homes to have one (Fig. 14.3). In the following 15 years penetration jumped to eighty percent. After 40 years of little interest in phones in the home, followed by another half-century of only a steady increase, this was a surprising change. The demand for new telephones went from around 300,000 per year to approximately a million.[34]

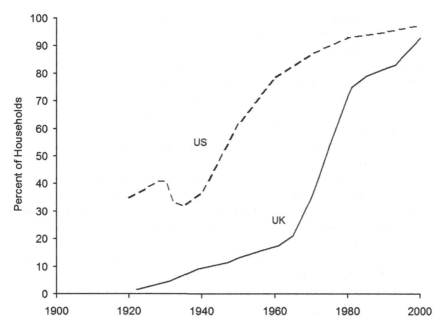

Fig. 14.3 Proportion of homes in the US and the UK with fixed telephones, 1900–2000. The Great Depression in the US is clearly seen, though it hardly affected the UK. Source: Author.[35]

The telephone changed from being something used only by professionals to an everyday item for both business and home. Increasing affluence was certainly a factor, and those used to having to depend on phone boxes to keep in touch with their loved ones could afford a home phone. In an accelerating world, particularly with the expansion of university education, people needed a way that was more instant than letters to communicate with families that were more spread around the country. It is difficult now to appreciate how easy it was to lose touch with people who weren't on the phone.

The ability to dial anywhere in the country was a huge incentive. Subscriber Trunk Dialing (STD) was the final solution to the problem of converting the dialing code into the routing information which had only been partially solved by the director system. More sophisticated technology had provided a solution to the complex code-changing problem.

The introduction of STD in 1958 and its rapid extension across the country during the 1960s, together with the charging of calls by their length, and cheap rate periods, was a great boon. Before that, for anywhere out of the caller's local area the call had to be connected by the operator who would break in every three minutes to see if you wanted to extend it. In the late 1960s and 1970s it also became possible to direct dial numbers abroad. By 1980, International Direct Dialing was available to over ninety percent of UK phone customers and they could call over 87 countries.[36]

A further development was the introduction of phones with push buttons rather than dials in 1976. The pulses needed by the Strowager-type exchanges were still generated, but produced electronically rather than by the controlled return of the dial. It was only in the 1990s with the introduction of all-electronic exchanges that the faster tone dialing became universal.

The other side of the coin was that to get a rapid expansion the Post Office had to be able to meet the demand. It had been unable to do this in the 1950s, although by the early 1960s it had got its backlog under control. The near disappearance of the waiting list provided a further incentive to the prospective user. It was only the freeing of its financial controls to allow the Post Office to borrow money, like other nationalized industries, that gave it the resources to install exchanges and cables at the required rate. From the late 1950s the capital demands of the telephone system doubled, and then doubled again.[37]

Once this expansion really got going, it became self-fulfilling. To make it worthwhile to have a phone, the intending renter needs other people like them to have one. As the system expands it reaches a point where their friends have one, which is an incentive to get one too. It is difficult to pin down exactly what starts this process, particularly as there was hardly any advertising until the Buzby campaign in 1976 designed to mitigate the effect of price rises. Undoubtedly the technical improvements made the system more user friendly, but in the end it is a social phenomenon which, once under way, has a life of its own.

After the initial rush, the growth slowed but then picked up again. Then it peaked as its child, the mobile cell phone, began to take over. Some 125 years after Bell's inspired prediction, it has come to pass. There is a telephone in virtually every home, and everyone talks to their friends without leaving it.

NOTES

1. Calney, M., The International Centennial Exhibition of 1876; or Why the British Started a World War, available at: larouchejapan.com/.../text/1876-Centennial-Exhibition.pdf; and also Report of the International Exhibition available at: http://archive.org/stream/internationalexh00unitrich#page/n37/mode/2up.
2. Bell was on stand N64 in the Main Hall although he also had a presence for Clarkes College upstairs in the East Galley. International Exhibition 1876, Official Catalogue available at: http://archive.org/stream/internationalex00commgoog#page/n32/mode/2up.
3. In the Main Hall Bell was surrounded by rivals and competitors. He had position N64 while the Western Union Telegraph Co. were at N63 and Elisha Gray was a short distance away on the Nave. According to the catalogue he was exhibiting a Harmonic Telegraph, but he actually had a telephone; DNB, Alexander Graham Bell.
4. US patent 161,739, Filed March 6, 1975, granted April 6.
5. Wheen, A., *Dot-Dash to Dot.com*, pp. 37–39.
6. US patent 174,465 Filed February 14, 1876, granted March 7.
7. Gray's application was only a "caveat" which is not an invention but the possibility of one. Thus, in any case Bell would have taken precedence; the argument was that he had had sight of Gray's application which included this.
8. Bell's articulating telephone, *Journal of the Society of Telegraph Engineers*, 519–525, 1876; also available at: http://earlyradiohistory.us/1876bel.htm.
9. Dom Pedro seemed to be on a one-man campaign to prove that monarchs weren't indolent.
10. Official Report by Sir William Thomson upon Mr. Graham Bell's Telephone, exhibited at the Centennial Exhibition at Philadelphia in 1876. Quoted in *Journal and Proceedings of the Royal Society of New South Wales*, Vol. XII, 4, 1978, available at: http://zapatopi.net/kelvin/papers/report_on_bells_telephone.html.

11. Farley, T., *Telephone History, IV. The Telephone Evolves*, available at: http://www.privateline. com/TelephoneHistory2A/Telehistory2A.htm.
12. Green, M., *The Nearly Men*, p. 18.
13. The Library of Congress, Special presentation: Emile Berliner, available at: http://memory.loc. gov/ammem/berlhtml/berlemil.html.
14. US Patent 463,569. (Though filed on June 4, 1877, disputes led to this only being granted in 1891. It was assigned to the American Bell Company.
15. Blake's patent: US patent 250,129.
16. Baldwin, F.G.C. The progress and potentialities of the telephone in the United Kingdom. *Journal IEE*, 61: 313, 24–30, 1922.
17. Pierce, J.R. The telephone and society in the past 100 years, available at: *thorngren.nu/.../ Pierce_John_R_1977_-The_Telephone_Past_100.pdf.*
18. The Telephone – The Connected Earth Project, available at: http://www.connected-earth.com/ Journeys/Firstgenerationtechnologies/Thetelephone/Developingthetelephone/index.htm.
19. his was a lot of money at a time when the average income was not a lot more than £1 per week.
20. Milne, G.J., British business and the telephone, 1878–1911. *Business History*, 49:2, 163–185.
21. Computed from data in Baldwin.
22. E.g., Milne.
23. Almon Strowger was an undertaker and it is said he invented the system because he thought he was losing business due to the local operator being the wife of his rival. Though he had some early success it took 20 years before it was generally accepted.
24. Freshwater, R. UK telephone history – Electrophone system, available at: http://www.britishtelephones.com/electrophone.htm.
25. US data from Baldwin and UK figure from data behind graph.
26. Data of total number of telephones to 1922 from Baldwin, then Post Office Annual Reports, and finally from 1982 CIA World Factbooks. Phones in homes from The Times May 16, 1931 (quoting the Assistant Postmaster-General), then Post Office Annual Reports, and the General Household Survey. (These are turned back into numbers using numbers of households interpolated from censuses.) Business phone numbers by subtraction of the other two.
27. 1919 data from the graphs in The Telephone Museum, The History of the Telephone System: available at: http://www.mkheritage.co.uk/ttm/telhistory2.html. 1939 and 1947 figures from Post Office commercial accounts 1947–1948.
28. Data from Freshwater, UK Telephone History, available at: http://www.britishtelephones.com/ histuk.htm.
29. For Birmingham codes as an example see: http://telephonesuk.co.uk/old_dialing_codes.htm.
30. Hill, D. The history of UK telecoms, Birchill's Telecom Museum available at: http://www. birchills.net/1930-1950-telephones/.
31. The Post Office spent £20 million in 1947–1948 on telephone investment – a huge sum for the time Post Office commercial accounts 1947–48.
32. Post Office commercial accounts 1947–48.
33. *The Observer*, August 5, 1962.
34. Post Office commercial accounts over this period.
35. UK data from same sources as previous graph but converted to percent of households using Household data; US data from: Historical Statistics of the United States, Colonial Times to 1970 (GPO, 1975), Part II, Chapter R, Series R 1-12, available at: http://www2.census.gov/ prod2/statcomp/documents/CT1970p2-05.pdf.
36. The Telephone – The Connected Earth Project, available at: http://www.connected-earth.com/ Journeys/Firstgenerationtechnologies/Thetelephone/Internationalnetworks/index.htm.
37. Post Office commercial accounts over this period.

15

Horseless Carriages: Road Vehicles

No other man-made device since the shields and lances of ancient knights fulfills a man's ego like an automobile.

Car magnate Lord Rootes

In 1900, more electric than gasoline (petrol) engined cars were sold in the United States (Fig. 15.1).[1] Even more steam-powered ones were sold, and the market was a battleground between the three types. The electric motors and systems were well developed from their use in trams and streetcars, and the storage batteries had also evolved over many years. Despite that, they were the weak point, giving the vehicle a range limited to 30 or 40 miles at low speeds before somewhere had to found to charge them at a time when electricity supplies were few and far between. At first, in a world used to constant changes of horses, the limited range didn't seem to matter too much, but it was to become the fatal flaw.

Steam cars had also been around for some time. As a power source it was the prime mover in industry and the railways. The problem was making it small and light enough to be suitable for road transport. By the turn of the century, these problems had largely been solved, but the steam car was still cumbersome, needed to be refilled with water every 30 or 40 miles, and required a considerable time to get up steam before being driven off.

The gasoline engine of the time was noisy and vibrated badly. Starting was difficult, with the driver having to go to the front of the car and swing the starting handle to turn the engine over. If he was lucky it started, otherwise he had to keep up the exhausting work until it condescended to fire. Once moving, there was also the difficulty of changing gear on the primitive gearboxes. These cars did have one great advantage: with a gas tank of a reasonable size they could go much greater distances without having to stop. In the end, this was what counted.

In Europe, the situation was different. Though the three types of engine were available, the internal combustion engine was already coming out on top. In Britain, the land of steam, it would have been expected that steam engines would have been in the ascendancy, but a series of ridiculous Acts of Parliament, no doubt supported by the railway lobby, had throttled them. A car could only be driven at 4 mph and then, following the 1865 Act, a

© Springer International Publishing AG 2018
J.B. Williams, *The Electric Century*, Springer Praxis Books, DOI 10.1007/978-3-319-51155-9_15

Fig. 15.1 A 1905 electric brougham—note that the driver is out in the cold and wet. Source: http://www.classicandperformancecar.com/front_website/gallery.php?id=486898

man with a red flag had to walk 60 yards in front of the vehicle. This drove everything but large traction engines off the roads.[2]

In 1896, this Act was repealed but the speed was still limited to 12 mph. To celebrate, the motoring lobby organized a car rally over the 54 miles from London to Brighton.[3] As the promoters were enthusiasts for gasoline-engined vehicles the distance was probably deliberately chosen to rule out electric cars and even put the steamers at a disadvantage. Not unexpectedly, the leading cars to arrive all had gasoline engines and were of foreign, largely French, manufacture.

The automobile was not invented by any particular person, but was the result of a long process of development where numerous inventors all made their contribution. Even the internal combustion engine was the work of a considerable number of people. Important amongst them were Frenchmen Jean-Joseph Etienne Lenoir and Alphonse Beau de Rochas, and the Germans Nicolaus Otto, Gottlieb Daimler, Karl Benz, and Wilhelm Maybach.

The first engines used gas as a fuel and then migrated to gasoline as it became available as a by-product of the refining of oil to obtain kerosene or paraffin for use in lamps. What linked all of these was the need to ignite the fuel at the correct moment to provide the power stroke of the engine. This was where electricity came in.

Some experimenters tried using a hot wire, sometimes curiously called the "catalytic ignition", but this used a lot of battery power and it was very difficult to control the moment at which ignition took place.[4] It didn't produce a satisfactory engine and was soon abandoned in favor of systems based on generating a spark.

The electrical industry had long known how to do this. A coil in series with a battery allowed a current to flow and the current to increase in the circuit. If this circuit was suddenly opened then the current tried to continue flowing in the coil and so was seeking somewhere to go. As a result, a large voltage was generated across the contact that had opened, and a spark ensued.

In the early stages, when batteries weren't very satisfactory, a small generator was substituted. This used a coil for one side and a permanent magnet for the other and so was known as a magneto. As the magneto was connected to the shaft of the engine its output varied with the engine speed. It was easy to make this adequate for all normal working conditions, but starting was often rather difficult as there was little voltage when turning slowly.

The disadvantage of these "low voltage" systems was that the opening contact had to be inside the cylinder of the engine. All sorts of ingenious mechanical arrangements were devised to achieve this, and the engines actually worked. However, the contact soon burnt away and the arrangement wasn't very reliable. A better system was needed.

The obvious thing to try was a Ruhmkorff coil which was basically an electric buzzer consisting of a coil which would open a contact. This would then close again under a spring. The result was that the device would keep switching on and off. It was a simple way to convert the direct current from a battery or magneto into alternating current that could then be stepped up to a high voltage using a transformer with a small number of primary turns and many on the secondary. In practice, the switching coil and the transformer could be combined into one device. The vibrating arms were known as "tremblers".

This worked slightly differently compared with the low voltage systems. First the spark started when the contacts on the engine shaft closed rather than opened but, more importantly, the spark didn't occur at the opening contacts which were now outside the cylinder and all that was needed there was a pair of contacts. These soon developed into the screw-in "sparking plug". This was rather more satisfactory, but the tremblers needed to be set up correctly and in practice gave a lot of trouble.

A way to eliminate the trembler was to return to the single spark system used in the low voltage arrangements but again use a transformer of high ratio to step up the voltage. Usually the coil and the transformer were combined into one ignition coil. This returned to producing the spark when the contacts in the low voltage circuit opened but, like the trembler system, it occurred away from the opening contacts and at the sparking plug in the cylinder of the engine.

At this point another American farm boy entered the story. He was Charles Franklin Kettering whose bad health and eyesight delayed him obtaining his degrees in mechanical and electrical engineering until he was 28 (Fig. 15.2).[5] Once he had graduated he went to work for the National Cash Register Company in Dayton, Ohio. His inventive contributions soon impressed his colleagues and in particular a young executive, Edward A. Deeds.

When Kettering discovered that the Cadillac automobile company in Detroit was dissatisfied with all the ignition systems, he began to work in his spare time to produce an improved one. This was so successful that Cadillac's chief engineer, Earnest Sweet, came to examine and road test the system. He spent eight hours driving the vehicle over the hills around Dayton and in all that time the ignition performed perfectly. A few weeks later, Kettering was called to Cadillac's headquarters in Detroit. There, head of Cadillac, Henry M. Leland, handed him a contract to supply 8000 of his systems.[6] Hastily Deeds, Kettering and his assistant, W.A. Chryst, set up a new company, Dayton Engineering Laboratories Company or "Delco", to fulfill the contract.

Kettering was to patent many variants of the ignition systems, but his aspirations went further.[7] His break came following a tragedy when a motorist tried to help a lady whose

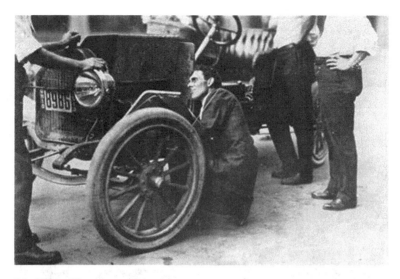

Fig. 15.2 Charles Kettering working on a car. Source: http://www.ohiohistorycentral.org/w/
File:Kettering,_Charles_F.jpg

automobile had stopped. As he was turning the starting handle, the vehicle backfired and
he was hit in the face, breaking his jaw. Though he was taken to a doctor, he died a few
weeks later. The incident concerned the senior people at Cadillac and they called on
Kettering to produce a reliable self-starting system.[8]

The arrangement he came up with involved using a motor/generator. The motor was
used via reduction gearing to start the vehicle, and then switched over to recharge the
battery. It was Kettering's skill that produced a reliable system, sufficient for Cadillac to
introduce it on their 1912 models. (They weren't entirely convinced as they still fitted a
magneto and supplied a starting handle, just in case.) Other manufacturers quickly followed
suit, and by the early 1920s virtually all automobiles were fitted with electric starters,
though gradually separate starter motors and generators became the standard as it was
simpler to implement.

This was the breakthrough for electric devices on vehicles. With a constant supply of
electricity from the storage battery that was kept charged by the generator or dynamo,
magnetos soon disappeared from automobiles and battery ignition became the norm.
Electric lights, instead of ones powered by acetylene gas, had been used on electric
vehicles, but now they were practical on gasoline-driven ones. Other auxiliaries such as
windshield wipers also became possible, and the way was opened to providing the horn
as well as myriad dials, gauges and warning lamps.

This, and the mass production methods of Henry Ford and his competitors in bringing
down prices (see Chap. 6), led to a great exuberance in America. The country loved the
automobile and bought it in phenomenal numbers. In 1919, there were 6.7 million vehicles
in the US, but a decade later this had rocketed to 27 million which meant that one in every
five people owned one.[9]

At this point, some 60% of households had a vehicle, vastly more than in other countries
but the financial crash in 1929 brought an almost immediate halt to this. Numbers didn't

Fig. 15.3 A motor omnibus from 1905. Source: *Dundee Courier*, September 2, 1905.

recover to this level until well after World War II. After that they began a much steadier rise, much like other countries but at a higher level.

In Britain, during the period before World War I automobiles were the prerogative of the rich not all of whom drove with due consideration for the other, slow-moving horse-drawn vehicles on the roads. As the vehicles gradually became more efficient the numbers on the roads steadily increased from 128 in 1903 when registration began to 132,000 in 1914.[10]

The improvements largely removed the drawbacks of gasoline-driven vehicles and electric and steam autos began to disappear. Electric cars hung on for a while—there were some 6000 electric and 4000 commercial vehicles registered in London in 1910.[11] Though the automobiles gradually disappeared, electric drives remained popular for local delivery vehicles which had to frequently start and stop. Electric milk floats for doorstep delivery were used well into the second half of the century. Steam remained for heavy delivery vehicles for a while, but gave way to diesel engines when they became available.

Even more extraordinary was the rise of the gasoline-driven bus (Fig. 15.3). They began to appear on the streets of London and other cities in competition to the horse-drawn buses in 1905. Over the next 7 years the 5000 horse buses (and their 18,000 horses) in London disappeared and were replaced by 2500 motor buses.[12] Though at first they were much criticized for their noise and vibration, particularly by house owners along their routes, they were very popular with ordinary people who couldn't dream of owning their own vehicle.

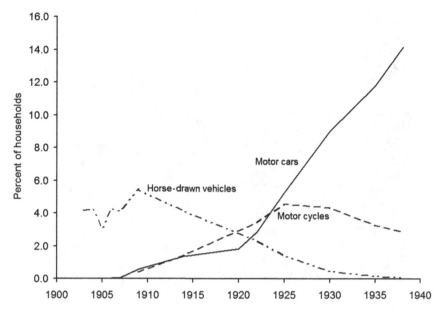

Fig. 15.4 The rise of motor cars and cycles meant that horse-drawn carriages disappeared by 1939. Source: Author.[13]

For those unable to afford an expensive automobile, a motor cycle was an attractive option. These had developed naturally as an extension of the strong cycle industry in Britain. Before World War I their numbers were keeping pace with cars, but shot up after the war as many young men obtained them. Their rise was probably faster than Fig. 15.4 suggests as there are no figures for wartime when the numbers almost certainly leveled out, and in 1919 115,000 motorbikes were sold.[14] They were very popular in the 1920s and for a number of years there were more on the roads than cars (Fig. 15.5).

Of course, what the aspiring motorist really wanted was an automobile, but the offerings of the motor industry had been concentrated on large, and hence expensive, models. Herbert Austin had set up his motor company in 1905. It had slowly progressed until the war when it expanded enormously on military work. Once into the postwar slump he was looking for ways to fill his factory and decided to produce a small automobile. This was partly in response to the new 1921 car tax which was based on the nominal horsepower of the engine, and so favored a smaller engined vehicle.

The Austin 7 was nominally of 7 horsepower, but though it had all the equipment that by now had become standard, it was small, reliable and cheap to buy and run. The 1924 model, for example, could reach 40 to 50 miles per hour and could go that distance on a gallon of fuel, all for $600 (£149).[15] Probably unwittingly, Austin had moved from catering to the luxury market to making ownership possible for the middle income

Fig. 15.5 Typical motor cycle, 1919. Source: Author.

ranks of society. Rapidly, other manufacturers such as Morris started to produce competing vehicles.

There has been considerable criticism that the manufacturers didn't take full advantage of Ford's production line methods which would have lowered prices and opened the market to a wider clientele. This ignores the very different factors between Britain and America. America had a higher standard of living, and a much larger potential market, so it was thus practical for Ford to take the route he did. In Britain, the numbers were much smaller and the market was more sensitive to the cost of maintaining the vehicle on the road than to its purchase price, which was largely financed by installment plans or hire purchase.[16] Under these conditions, the manufacturers made the right decisions and the market grew quite fast enough for them.

For those able to afford them, automobiles brought a new freedom. Many people, particularly the young and single, took to the roads to explore their own country (Fig. 15.6). With the roads beginning to be tarred over, and with still not a lot of traffic on them, the interwar years were the golden age for those wanting to wander and discover their country's hidden treasures. Guides such as those published by Shell, and books like H.V. Morton's *In Search of England*, encouraged new motorists to take to the roads.

The non-drivers were catered for by the rise of, first, of the charabanc—the primitive type of coach with banks of seats and (at first) no roof. This metamorphosed into the real coach which could take the Sunday School on its outing or the mill workers to the seaside. Coach travel was simpler and more flexible than the train and could take you right to the door. People who had never ventured outside their town or village could see how the rest of the country lived.

Fig. 15.6   Intrepid travelers with their baby Austin. Source: Author.

For day-to-day transport, the motor bus continued its rise. Streetcars or trams reached their peak in the mid-1920s, but in 1928 Manchester replaced their streetcars with motor buses. From then on, tram rails were ripped up rather than put down as the motor bus steadily took over. By 1932, buses were carrying more passengers than trams.[17] In the countryside, rural buses were bringing a new mobility to those beyond the reach of the railways.

Towns and cities, which had started to expand along electric railway and streetcar routes, now spread even further, as people could more easily get to them by bus and, increasingly, the automobile. Deliveries could be made using the extensive range of motor vans and trucks to transport goods from the store or the railroad station to the house door, wherever it was.

All this increased traffic needed to be controlled, the high rate of accidents reduced. Traffic signals were introduced at busy junctions. The first were simple semaphore arms, but electric lights were much more satisfactory. As it is simple to switch lamps on and off, they were soon made automatic and gained the nickname "robots". Though this was soon replaced by the prosaic "traffic light", it still lives on in South Africa.

In the vehicle itself, instead of having to put a hand out of the window to indicate that the vehicle was turning, the "trafficator" was added. This was a small lit arm on each side that projected behind the front doors when activated. It was operated by a solenoid which was activated by a switch, usually on the steering column. Later, of course, it was replaced with the flashing amber lamps on the corners of the vehicle that are still with us.

In Britain, the energetic interwar Minster of Transport, Leslie Hore-Belisha, revised the Highway Code to introduce more traffic control measures. He is perhaps best known for his "zebra crossings" to allow pedestrians to cross the road in safety.[18] This consisted of a striped "zebra" marking on the road and on each pavement a pole topped with a flashing

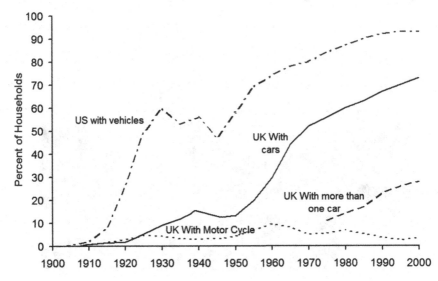

Fig. 15.7 Households with cars and motor cycles, 1900–2000. Source: Author.[19]

amber ball –the Belisha beacon. Once again it was the ease of switching electric lamps that made this possible. They have now mostly been replaced by the more sophisticated pelican crossings, but they can still be found in out-of-the-way places.

By 1939 the number of motor vehicles on the roads in Britain had grown to some three million. Approximately two million of these were automobiles, but this still only represented little more than 15% of households having a vehicle.[20] The war, while it brought improvements in the engineering as in America, caused a decline in the numbers of automobiles which weren't to recover to the prewar level until around 1950 (Fig. 15.7). The exception to the trend was motorbikes, where ownership rose markedly once the war was over and continued to rise until around 1960. The total number of vehicles of all sorts, which had remained static during the war, began to climb steadily upwards.

The 1950 and 1960s were a time of general prosperity and with that went the desire to own a vehicle. Soon the numbers even outside America were large enough for real mass production, leading to a virtuous circle of increasing numbers and decreasing real prices.

Gradually the specification of the automobiles improved, with many more auxiliaries and extras, most of them electrical. Belt-driven fans were replaced by electric ones and electronic management systems squeezed ever higher performance out of the smaller engines. Electric motors opened and closed the windows and waggled the mirrors. Door locks were operated electrically which led to central locking, with "keys'" that could operate remotely at the press of a button. And, of course, there were entertainment systems — radios, tape and CD players.

All this extra electrical equipment threatened to overload the dynamos so AC alternators were fitted once it became practical to use electronic rectification and control to charge batteries from them. The automobile, which had began as a pure mechanical system, now contained a vast amount of electrical and electronic equipment, let alone the miles of wire. It was controlled by electronics and much of its sophistication came from the electrical systems.

At first sight, a diesel engine would seem to be free of electrics. As it uses compression to ignite the fuel, it doesn't need that electric spark. In practice, a diesel-powered vehicle has nearly as much electrical equipment as a gasoline- or petrol-engined one. It only lacks the ignition system, but otherwise contains everything else. The fuel injection needs careful electronic control so as not to cover the following vehicles in soot.

Late in the century there was a resurgence of interest in electric vehicles around the world. Despite some progress in storage batteries there was nowhere near the improvement seen in other fields. The problem of battery capacity and, hence, range of the vehicle still remains. Various hybrid schemes using electric motors and smaller engines were tried, but until a shortage and therefore a high price for fossil fuels applies more pressure, they are unlikely to become mainstream.

By the end of the century, the automobile had become ubiquitous with some 23 million on the roads in Britain alone, which was nearly as many as the number of households.[21] Although it appears that virtually every family had one, it wasn't like that. In practice, it didn't quite reach three-quarters. The discrepancy is explained by the fact that 28% of households had more than one car, sometimes three or even more.[22] In America, there were some 126 million vehicles or nearly twice as many as the number of households.[23] This showed an even greater propensity for multiple vehicles as there was still a few percent of households without one at all.

During the twentieth century the automobile grew from being a rich man's toy to becoming an essential part of everyday life. It has shaped society, altering where people live and work. It has defined life styles and even jobs, as well as leisure. At first sight it is a mechanical device, but without the electric spark it wouldn't go and, lacking all those electrical extras, it would still be a crude and uncomfortable jalopy.

NOTES

1. The figures were 1681 (40.1%) steam, 1575 (37.6%) electric and 936 (22.3%) gasoline. This is from multiple sources, e.g., Carhistory 4 U, History of Motor Car / Automobile Production 1900–2003, available at: http://www.carhistory4u.com/the-last-100-years/car-production.
2. A good write-up is in the *London Daily News* of November 14, 1900 in an article on the history of the motor car.
3. *The Guardian*, November 16, 1896.
4. Warr, J.W. The electric ignition of internal combustion engines. *JIEE*, 44:199, 134, 1910.
5. Jeffries, Z. Charles Franklin Kettering 1876–1958, A biographical memoir. National Academy of Sciences 1960, available at: http://www.nasonline.org/publications/biographical-memoirs/memoir-pdfs/kettering-charles.pdf.
6. Barach, J. Motor era, available at: http://www.motorera.com/history/hist05.htm.
7. E.g., US patents 1,162,073, 1,163,092, 1,167,762.
8. Barach.

9. Digital History, The consumer economy and mass entertainment, available at: http://www.digitalhistory.uh.edu/disp_textbook.cfm?smtID=2&psid=3396.
10. See the Postmaster General's reports on the Post Office for these years.
11. Electric Vehicle Association, 1991, quoted in C. Ivory and A. Genus, Symbolic consumption, signification and the 'lockout' of electric cars, 1885–1914. *Business History*, 52: 7, 1107–1122, 2010.
12. *The Times*, Success of the motor omnibus, February 27, 1912.
13. Data from: Postmaster General's reports on the Post Office from 1904 to 1915; Board of Trade Statistical department reports 1931 and 1939; Department of Transport, statistics table veh0103, available at: https://www.gov.uk/government/statistical-data-sets/veh01-vehicles-registered-for-the-first-time; household numbers extrapolated from censuses.
14. Pugh, M. *We Danced All Night*, p. 244.
15. Vintage Austin Services, Engineering services for vintage Austin sevens, available at: http://www.vintageaustinservices.co.uk/.
16. Bowden, S. Demand and supply constraints in the interwar UK car industry: Did the manufacturers get it right? *Business History*, 33:2, 241–267.
17. Taylor, A.J.P. *English History 1914–1945*, pp. 303–304.
18. Pugh, M. *We Danced All Night*, p. 256.
19. Data from: UK: Postmaster General's reports on the Post Office from 1904 to 1915; Board of Trade Statistical department reports 1931 and 1939; Department of Transport statistics table veh0103, General Household Survey, Table 6 Consumer Durables; household numbers extrapolated from censuses; US, McGrath, R., The pace of technology is speeding up. Harvard Business Review, November 25, 2013, available at: https://hbr.org/2013/11/the-pace-of-technology-adoption-is-speeding-up, originally from: http://www.nytimes.com/2008/02/10/opinion/10cox.html? ex=1360299600&en=9ef4be7de32e4b53&ei=5090&partner=rssuserland&emc=rss&pagewanted=all&_r=1&_ga=1.125098697.291417130.1408289727&.
20. Department of Transport, statistics table veh0103.
21. Vehicle numbers from Department of Transport statistics table veh0103, household numbers extrapolated from censuses.
22. General Household Survey, Table 6 Consumer Durables.
23. Tang, S. History of the automobile: Ownership per household in US, available at: https://en.wikibooks.org/wiki/Transportation_Deployment_Casebook/History_of_the_Automobile:_Ownership_per_Household_in_U.S.

16

Too Cheap to Meter? Nuclear Power and Beyond

Our children will enjoy in their homes electrical energy too cheap to meter.

Lewis L. Strauss, chairman of the U.S. Atomic Energy Commission, 1954

With most homes now connected for electricity, and industry and commerce increasingly dependent on it, governments began to take a greater interest. They were concerned about the natural monopoly and controlling financial excesses, but also about the reliability and service provided by the utilities. In America, land of free enterprise, this took the form of heavy regulation of the private companies. In many other countries, although the utility appeared to be a private company, it was mostly owned by the state.

In Britain, the part-and-part system for generation and distribution set up before the war worked reasonably well. However, the much more intractable problem of local distribution had been left to fester. There was a growing consensus that something needed to be done, but the governments of the 1930s were largely dominated by the Conservatives who were too divided to apply any solution. The simple approach of persuading the various authorities to combine foundered, as ever, on the mistrust, and in some cases outright hatred, that the municipal and company organizations had for each other.

One man, Herbert Morrison, was quite clear in his mind what should be done. He had a long involvement with the electricity industry and knew the problems. His was a simple solution: nationalize the lot. When the Labour Party won the election of 1945 by a landslide, he was given the job of coordinating the legislative program for nationalizing the various industries that they had earmarked, among them electricity.

The new arrangement came into force on the propitious date of April 1, 1948. A new British Electricity Authority (BEA) operated the generation and bulk distribution, with 14 Area Boards to sell the electricity to customers. There had been some opposition, particularly from Wade Hayes of Edmundsons and, of course, Balfour Beatty, but generally the logic of the situation was so clear that it was generally accepted. In any case, many of the companies were now liable to be compulsorily purchased by the local municipality, because of the 42-year time bomb that Joseph Chamberlain had planted back in the previous century. Maybe nationalization, with full compensation, was just as acceptable an option.

© Springer International Publishing AG 2018
J.B. Williams, *The Electric Century*, Springer Praxis Books, DOI 10.1007/978-3-319-51155-9_16

147

It was a tremendous undertaking. There were 200 companies and 369 municipal undertakings to be taken over, together with the Central Electricity Board and the 297 generating stations that were owned by these bodies.[1] In practice, it was even more complicated as many of these organizations owned others and the numbers involved were huge. The new South Western Electricity Board, for example, took over some 80 organizations although the number of actual owners was smaller.[2] They varied enormously in size from large companies like NESCo and corporations like Birmingham, to tiny undertakings in very small towns with a handful of customers.

The generators were also a very mixed bag. Of the 297 generating stations, only a little over half, 155, had a rating of more than 10 MW, and most of them had been used as selected stations by the Central Electricity Board. Of the rest, half of those, 73, couldn't even produce 1 MW, with the lowest at New Quay in Wales only managing 35 KW.[3] Added to this, less than half the 9.3 million customers were supplied with AC at the so-called 230 v standard voltage, and 12% were still on DC.[4] Clearly, there needed to be a sort out.

The new body inherited a whole raft of problems. The most important was the insufficient amount of generating capacity and that the CEB's forward program was hopelessly over-optimistic and far more than was really needed. Even when a more reasonable plan was instituted there were considerable problems due to the shortage of steel which caused a bottleneck in the construction of boilers and site work. Slowly but surely, the new British Electricity Authority got a grip of the situation, but despite their best efforts, the long lead times for everything meant that they were desperately short of generating plant into the 1950s and even then it was only just about adequate, with power cuts following sudden surges in demand or plant breakdown (Fig. 16.1).

It only required a cold winter or two, as occurred in the early 1960s, for the customers to experience enforced cuts rather than the occasional "shedding". After that, things started to improve, as the rate of introduction of new generating capacity was increasing and that of growth of maximum load started to decline. This was not due to a decreased demand (if anything, it accelerated), but that the "load factor" was improving. Instead of the sharp peaks morning and evening, the demand was more even throughout the day. The ratio of day peak to night trough halved from 3.4 in 1947/8 to 1.7 in 1971/2.[5]

The CEB took a certain amount of criticism for the excess plant in the 1970s, but when the position is examined, there was no clue in the 1960s, when the decisions had to be made, that the maximum demand would climb more slowly. One factor in this was the increased use of night storage heaters as a form of central heating. The industry encouraged these with cheaper tariffs at night to try and even the load curve. The heaters could (at least in theory) continue to give out heat throughout the day.

At the time of nationalization around a quarter of homes were still not wired for electricity, mostly in poor urban areas and out in the countryside. However, an extensive housing drive in the 1950s regenerated many urban areas, and the 1953 plan to connect 85% of the farms within a decade was completed 18 months ahead of schedule.[6] Using wooden poles and simplified equipment kept the costs down, and rural users were soon using much more electricity than most households as they could install milking machines, better lighting, pumps for watering and so on. The financial losses turned out to be lower than expected.[7]

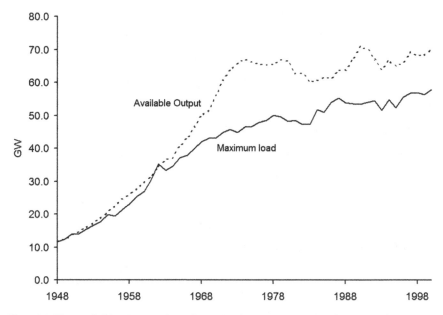

Fig. 16.1 The available output and maximum load demanded, 1948–1998. In 1950 and 1962 the system was unable to supply the demand. Source: Author.[8]

At the end of the first 10 years of nationalization, less than 10% of households still didn't have a supply.[9] By 1965, most of the UK was wired, though in the north of Scotland and Northern Ireland connection rates were well behind the rest of the UK. In England and Wales around 98% of households were connected by this date.[10]

Most countries now found that, as consumption continued to increase (and Fig. 16.2 for the UK is fairly typical), the distribution networks were no longer adequate. The usual solution was to add another "Supergrid" running at an even higher voltage to move the increased amounts of power. It meant that power stations could be sited where it was convenient, for example near coal mines or along rivers where cooling water was available, and the power transferred to where it was needed. A spectacular piece of siting in Britain was at Rugeley in Staffordshire, where the power station was on top of the coal mine with a conveyor delivering fuel directly from the pithead to the boilers.

Eventually 4900 kilometers of Supergrid lines were added to the original 132 kV British Grid. They had begun as 275 kV which was sufficient until the mid 1960s. The system had been designed to be up-rated to 400 kV after that date. By 1977, the Grid was mostly 400 kV with local networks at 275 kV, finally feeding 132 kV and the lower voltage distribution lines. The whole system had been modified but the concept of the Grid was still there. Britain had a much more integrated Grid than other countries.

The rising demand in all countries meant that increased generation capacity was needed. At first sight this is merely a matter of building more stations.

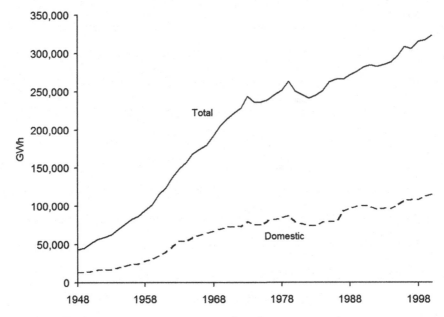

Fig. 16.2 Electricity consumption in the UK, 1948–1998. The growth was interrupted by the fuel crisis in 1974 and the slump of the early 1980s. The domestic consumption remained around a third of the total. Source: Author.[11]

However, it had long been known that increasing the size of the generators brought benefits. There were two advantages: they were more efficient in that they burnt less fuel for the same output, but also the capital cost per MW of output was lower. Thus, the trend was to build fewer stations but with generators of larger sizes. Figure 16.3 applies to the UK, but is fairly typical of what was happening elsewhere.

After the World War II, there was a very conservative policy of using 30 and 60 MW sets, but soon there was a jump to 100 MW and 120 MW sets. Though there were some difficulties as each new generation was introduced, the utilities pressed on with ever larger generator sizes. By 1966, this had reached 500 MW which remained the standard for a while before rising to 660 MW in the 1970s. After this, progress largely halted due to many factors, not least of which was the pure difficulty of moving the enormous turbines and generator parts from the manufacturer to the station.

The high stakes of the game can be seen in that a 2000 MW station (four 500 MW generators) in the 1980s would have an annual fuel bill of around $300 million, so squeezing the last bit of efficiency out of the system was vital. However, a breakdown meant that the replacement generating cost of a large turbine generator could be between $60,000 and $200,000 per day.[12]

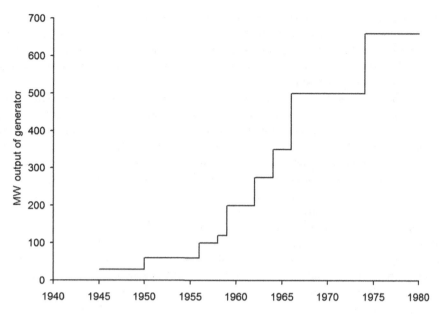

Fig. 16.3 Increasing size of generators and the approximate date of their introduction, 1940–1980. Source: Author.[13]

All this technology still largely rested on coal being fed into furnaces. Despite the improvements in efficiency, the rapidly increasing demand for electricity needed an ever-increasing quantity of coal. Some countries had considerable resources of hydroelectricity, before long all the best sites had been used. Oil was considered but was generally more expensive than coal, though gas was a partial alternative.

There were increasing doubts in many countries as to whether the coal industry could cope, and governments, particularly in Britain, didn't want to be dependent on the miners. Thus the search was on for some other more efficient fuel.

In 1956, the Suez Canal was closed which meant there was no easy route to carry supplies across Europe and the Middle East. The British government went into panic as the oil supply was severely disrupted. Though it turned out to be a relatively short-term problem, the Suez affair concentrated the search for another source of primary power from which to generate electricity. This was many years before the discovery of gas in the North Sea, and so the options were strictly limited.

World War II was finally ended with two atomic weapons dropped on Japan. One used enriched uranium and the other plutonium as the fissile material. The US, Russia and Britain moved to develop these weapons and built atomic reactors to produce the plutonium. A large amount of heat is generated as the reactor is in a sense a carefully controlled explosion. As the size of these reactors increased, the need to get rid of the heat became clear.

An obvious solution was to use it to raise steam, and from that generate electricity. In 1954, the Russians connected their station at Obninsk, with a net electrical output of 5 MW, to the power grid.[14] Two years later, Calder Hall 1 in Britain, with a net electrical output of

50 MW, was connected to the national Grid. In the US, a nuclear power plant at Shipping port, Pennsylvania, was generating 60 MW by the end of the following year. Any of these could claim to be the first commercial nuclear power station, depending on how this is defined. Undoubtedly the first two were primarily for plutonium production, with the power generation as a byproduct.

In Britain, desperate for more power, the opening of Calder Hall by the Queen was accompanied by a lot of hype.[15] Though the reactor was intended to produce plutonium, it was also designed as a prototype of civil nuclear power stations. Almost immediately, on its successful operation, two commercial stations were ordered to be built at Berkeley on the Severn and Sizewell on the Suffolk coast. This was in addition to the Atomic Energy Authority's program for more reactors at Calder Hall and also at Chapel Cross in south Scotland.

It was around this time that there was a lot of loose talk about cheap electricity, including the quote at the top of this chapter, which was repeated many times over and became associated with nuclear electricity in particular. It was based on an erroneous idea and touted by people who should have known better. The uranium fuel used in the reactor was much cheaper than the coal needed to generate the same amount of power, and so the uninitiated jumped on this as meaning that the power would be cheap.

Unfortunately, it doesn't work like that. A very significant part of the cost of the electricity is the recovery of the capital cost of building the station in the first place. It is obvious that a nuclear reactor is a far more complex, and hence expensive, item to build than a conventional boiler for coal or oil firing. In the event, the station costs were three or four times that of a coal-fired station and so, on reasonable assumptions, the electricity cost was higher. There was some hope that as more stations were built, this would fall.

Despite suggestions that before long only nuclear stations would be built, in Britain there were only nine commercial stations constructed of the Magnox type based on the Calder Hall reactor.[16] Even so, by 1967 Britain had generated more power from nuclear stations than the rest of the world put together.[17] Though they were only supposed to run for around 20 years, the least successful generated for 26 years and the others for anything up to 45 years. The last to be built, Wylfa, which opened in 1971, finally shut down in 2015. With these extended lifetimes and generally reasonable reliability, the economics are more difficult to access. They have certainly saved many millions of tons of coal or oil.

In 1957, one of the very early reactors at the Windscale atomic fuel plant caught fire. Despite the heroism of some of the staff, there was a release of contaminants into the local area before it could be extinguished. This meant, for example, that milk could not be used for a few days until the radioactive components had died away. The incident shook much of the confidence in the safety of reactors. The particular core was one that had been hastily assembled to produce nuclear fuel after the war and didn't compare even with Calder Hall and particularly with the commercial Magnox reactors.

As a result, the British nuclear regulator had a severe attitude to safety. The reactors had extensive instrumentation measuring temperature, pressures, the flow of coolant and nuclear processes, all of which were fed to a Guard Line. This worked as a sort of chain; all the links had to be intact or the control rods would automatically be released and drop into the reactor, shutting it down. The instruments looked at the levels and even rate of change of the parameters to ensure that they weren't too great. The result was a very safe system, proof against meddling by the operators. The result was that there was never a nuclear incident on any of the commercial stations.

Despite all the nuclear station building, the bulk of electricity was still generated from coal, and by the 1970s the miners were becoming restive. A strike in January 1972 was soon felt in the power stations, and in power cuts for consumers. But by late February the miners had been bought off with a large pay rise and everything returned to normal. Then, in October 1973 Egypt and Syria attacked Israel in the Yom Kippur war over unfinished business from earlier conflicts. The US supported Israel which led the Arab oil states to increase oil prices by 70% while at the same time reducing supplies. The effect was soon to produce a crisis in Britain and other western countries.

In November, the National Union of Mineworkers (NUM), who had learnt their strength from the 1972 strike, saw their opportunity for another pay claim and instituted an over-time ban. By mid December the situation was becoming critical. The power stations were receiving 40% less coal than normal, stocks were falling rapidly, and attempts to substitute oil were affected by the shortages. Edward Heath, the Prime Minister, was exhorting everyone to save as much electricity as possible, with ministers even urging people to share baths and brush their teeth in the dark. To try to enforce the savings the television stations were shut down at 10.30 p.m.[18]

Worse was to come; from the end of the year commerce and industry would be subjected to a three-day week. Domestic customers were warned that if consumption did not fall they would be subjected to enforced cuts.[19] In the event, that also became necessary when the miners went on strike. One and a half million people became unemployed.[20] It would take a general election and change of government before the situation was sorted out.

The disruption was immense. All aspects of life were affected and it was difficult for almost everything to function in the times when the power was turned off. Though the cuts were only for specific periods it was an object lesson in how dependent the whole country had become on electric power.

The government were thus naturally very keen on nuclear power. Throughout the crisis the Magnox stations had supported a large amount of the base load. In the mid to late 1960s a start had been made on building the first five of a new generation of reactors. These were Advanced Gas-cooled Reactors (AGR), using higher temperatures and so able to generate electricity more efficiently. They still used carbon dioxide to remove the heat.

Unfortunately, the first contract, for Dungeness B, was given to the weakest of the construction consortia. It ran into all manner of difficulties which meant that the plant was not fully operational until 1985. The next two stations, though started 2 years later, were on power in 1976. Two further stations were ordered, and after that only one more of a different design. The enthusiasm for nuclear power had died.

In the US, Russia and France programs for nuclear power stations were also started. These were on much larger scales than the British effort. They were nearly all based on a water-cooled, rather than gas-cooled, design which was physically smaller and hence cheaper to build. Like the British commercial reactors, they had much larger outputs of 1000 MW or more. The US was to build more than 100; France built more than 50 and generated more than three-quarters of its electricity from them.

The Soviet Union built a large, but unknown, number, partly on its own territory but also in its satellite states across Eastern Europe. Unfortunately, though skilled people were operating them, their safety systems left something to be desired. For a period in the 1970s, nuclear power appeared to be the future of electricity generation. Only small amounts of uranium were required and the fuel rods only needed to be changed infrequently. There was no shoveling of large amounts of coal. They appeared clean and modern.

However, there was a concern was over safety. Accidents at Three Mile Island in the US in 1979, and at Chernobyl in the Ukraine in 1986, made everyone fearful. The British gas-cooled reactors were more expensive to build than some of these other designs, but they had one great advantage. Because gas is less efficient at removing heat than water, the core is much larger (hence the greater cost). However, if something goes wrong the much larger core is able to absorb the heat whereas with a water-cooled one loss of the coolant is catastrophic.

It is thus much more difficult to ensure that a water-cooled reactor is completely safe. The problems at Fukushima in Japan have underlined this. Here the tsunami overwhelmed the defenses and the reactors shut down, but with the loss of power it was necessary to start standby pumps to keep the coolant flowing. Foolishly, these had been sited at the lowest level and so were flooded as well. Had they been mounted higher in the station they should have been able to safely cool the cores. The result of these various incidents is that no one wants nuclear stations any more. In the meantime, the existing reactors soldier on.

In 1990, the British Conservative government privatized the electricity industry. Most of the hard work building up the infrastructure had been done, and there was an adequate margin of plant. Demand was rising less rapidly and, particularly, the maximum load was tending to level out. It was thus an easy ride for those taking over the assets. While the Area Boards were turned into companies, the National Grid was made a separate organization owned by the regional companies. The CEGB, responsible for generation, was broken up into three companies.

Though there was a regulator who was supposed to control what was going on, the central planning to ensure that there was an adequate amount of generation plant was gone. There have been considerable changes of ownership of the various bodies, often to foreign state-owned organizations. It has been generally profitable for them and their shareholders, but the domestic customer has lost out.[21] The benefits of "improved efficiency" in the private sector have proved illusory.

Though they had been around for some time, in the 1980s and 1990s there was a sudden enthusiasm for gas-fired stations. Particularly in Britain after privatization, nearly all the power stations that were built were gas-fired. The majority are combined cycle gas turbine stations (CCGT).[22] These use one or more gas turbines (essentially jet engines that run on gas), the exhausts of which are fed into a heat exchanger. This waste heat is used to raise steam which drives a further steam turbine. The overall efficiency of these stations is high and they are relatively cheap and quick to build.

In America, with waste gas from oil wells, and then gas produced by fracking, gas-fired stations seemed ideal. However, the gas turbines can't really withstand continuous running and cracks in the turbine blades became a common problem. Many of the stations spent considerable periods out of service for overhauls.

In Britain, with a plentiful supply of gas from the North Sea, they were very popular, but as the gas runs out and must be imported, the use of these stations doesn't seem quite so clever. The strain on the foreign currency is borne by the country, but the companies have no interest in restraining this. It is only the rise in the price of gas that is driving them back towards coal.

With the coming of the new century, there has been a considerable move towards renewable sources, particularly wind turbines, and solar panels. Compared with conventional power stations their output is very small, so it requires large number of them. It takes time to find suitable sites and a long time for sufficient plant to be installed to change the mix markedly. The promise of wave and tidal power has not yet been fulfilled.

The enormous change over the century can be seen in the rise in the annual electricity supplied. In Britain, this rose from 180 to 357,266 GWh, or nearly 2000 times.[23] This is equivalent to about 6000 kWh for every person in the country, which compares with the 4 kWh in 1900 for a smaller population.[24] In America, the electricity supplied only rose by less than 1000 times due to its earlier start.[25]

Over the course of the twentieth century, electricity changed from a curiosity to something essential to everyday life.

NOTES

1. Hannah, L. *Engineers, Managers and Politicians*, p. 7.
2. Derived from: South Western Electricity History Society, The Archives, available at: http://www.swehs.co.uk/tactive/sparkhome.php.
3. Calculated from: spreadsheet DECC Publications, 1948 Power Stations in Britain, available at: http://webarchive.nationalarchives.gov.uk/20110825140104/decc.gov.uk/publications/basket.aspx?filetype=4&filepath=statistics%2Fpublications%2Fenergytrends%2F1_2010021214364 1_e_%40%40_1948powerstations.xls.
4. Calculated from Table 2, Electricity supply, distribution and installation. *JIEE*, Part I: General, 91: 39, 104–114, 1944.
5. Milne, A.G. Distribution of electricity, *IEE Electronics and Power*, 19:18, 440–444, 1973.
6. Electricity Council, *Electricity Supply in the UK: A Chronology*, c. 1987.
7. Booth, E.S. The electricity supply industry – yesterday, today and tomorrow. *Proceedings of the IEE*, 124:1, 1977.
8. Data from Department of Energy and Climate Change, Historical electricity data 1920–2010 available at: https://www.gov.uk/government/statistical-data-sets/historical-electricity-data-1920-to-2011.
9. Hannah, L. *Engineers, Managers and Politicians*, p. 71.
10. Bowden, S. and Offer, A. Household appliances and the use of time. *Economic History Review*, XLVII, 4(1994)5, 725–748. These data are a little suspect as the numbers vary too much around this date.
11. Data from Department of Energy and Climate Change, Historical Electricity Data 1920–2010, available at: https://www.gov.uk/government/statistical-data-sets/historical-electricity-data-1920-to-2011; and Digest of UK Energy Statistics (DUKES) 5.1.2 Electricity supply, availability and consumption, available at: https://www.gov.uk/government/publications/electricity-chapter-5-digest-of-united-kingdom-energy-statistics-dukes.

12. Hawley, R. and Marlow, B.A. Efficiency of steam turbine generators for central power stations. *IEE Electronics and Power*, 30:1, 23–27, 1984.
13. Data from Hannah, L., Engineers, Managers and Politicians, p. 114, and Booth.
14. European Nuclear Society, Nuclear power plants worldwide, available at: https://www.euro-nuclear.org/info/encyclopedia/n/nuclear-power-plant-world-wide.htm.
15. E.g., *The Times*, October 17, 1956.
16. Plowden, Sir E. The second Industrial Revolution. *The Times*, October 17, 1956; The name Magnox came from the magnesium oxide material which was used for the fuel cans in these reactors.
17. Harrison, B. *Seeking a Role: The United Kingdom 1951–1970*, p. 313.
18. Marr, A. *A History of Modern Britain*, p. 340.
19. *The Times*, December 13, 1973.
20. The National Archives, British Economics and Trade Union politics 1973–1974, available at: http://www.nationalarchives.gov.uk/releases/2005/nyo/politics.htm.
21. Domah, P. and Pollitt, M.G. The restructuring and privatisation of electricity distribution and supply businesses in England and Wales: A social cost–benefit analysis. *Fiscal Studies*, 22:1107–146, 2001.
22. DTI DUKES 5.11, Power Stations in the United Kingdom, available at: http://webarchive.nationalarchives.gov.uk/20070603164510/http://www.dti.gov.uk/energy/inform/energy_stats/electricity/index.shtml.
23. Department of Energy and Climate Change, Historical Electricity Data 1920–2010 available at: https://www.gov.uk/government/statistical-data-sets/historical-electricity-data-1920-to-2011.
24. Hannah, L. *Electricity Before Nationalisation*, p. 427.
25. Ayres, R.U., Ayres, L.W. and Pokrovsky, V. On the efficiency of US electricity usage since 1900. *Energy*, 30:7, 1092–1145, 2005.

17

Keeping it Fresh: Fridges and Freezers

The household refrigerator changed the way people ate and socially affected the household.

Bern Nagengast, refrigeration historian

On August 19, 1923 seventy-five-year-old Mrs. Elizabeth Clackstone of Hull, England, opened a tin of best-quality salmon and fed it to her family. The following morning she ate what was left over. This was a foolish thing to do because within a few days she was dead.[1] This was just one example of the many incidents that resulted from eating contaminated food. It was not well understood how rapidly bacteria could multiply in food that was kept at room temperature.

The grocer in the above case said Mrs. Clackstone should have covered it with butter, while the doctor thought that flies were the problem. Even as late as this date, many people didn't understand the risks presented by keeping food lying around. In fact, only a few men of science understood the mechanism that caused food to become lethal.

It had long been known that the life of fresh food could be prolonged by keeping it at a reduced temperature. Pantries were often kept damp and had cold slate shelves. Many stately homes had an underground ice house where ice, harvested in the winter, was stored to preserve foods for as long as possible. Particularly in America a whole industry grew up supplying ice which could be put into an "ice box" insulated cabinet to keep food cool.

Throughout the nineteenth century the hunt was on for an artificial way of making ice or keeping the cabinet cool. The starting point was the evaporation of a liquid. It is well demonstrated by the old physics lab experiment: A small beaker of ether is placed on the bench in a little puddle of water. Air is blown into the ether via a tube so that it rapidly evaporates. When an attempt is made to lift the beaker it is found to be stuck to the bench because the water underneath has frozen.

In evaporating, the ether needs a considerable amount of heat which it extracts from its surroundings, hence freezing the water. Early in the nineteenth century there was a realization that this could be used as the basis of a cooling machine and many attempts were made, but it wasn't until the 1870s that satisfactory systems were produced.

Probably the first was made by the German professor, Carl von Linde.[2] He began by using methyl ether as the refrigerant, but rapidly turned to ammonia. This was made to evaporate by lowering the pressure with a pump and, as it evaporated, it extracted the heat, causing the desired cooling. On the other side the vapor was pumped into cooling coils where it turned back into liquid. This was returned through a valve to the cooling side.

It was a simple, but not a small, device. It was powered by a steam engine and the first customers were the lager breweries who needed low temperatures for their process, and this allowed them to brew the whole year round. Later, in America, the use of such devices spread to meat packing, and it was refrigeration that made possible the shipping of meat from Australia and New Zealand right around the world to the UK. Thus the country was able to feed its growing urban population which was outgrowing the capability of local agriculture to support it.

Once electric motors were available they soon replaced the steam engines as they were more reliable and convenient. Despite this, the refrigeration systems were only used in commercial premises as they were far too cumbersome to consider in a domestic situation. Even so, they had a considerable impact in improving the food quality as it could now be transported and stored without deteriorating, which was particularly important in America with its much greater distances.

During the first two decades of the twentieth century in the US there were numerous attempts made to produce domestic refrigerators. The approach was usually to cut a hole in the top of an existing ice box and install a cooling unit.[3] Despite this short cut, all these systems failed due to a combination of technical problems and the potential market not being particularly receptive. Also, despite the US being ahead in domestic electrical connections, there were still few homes with power.

Three technical advances changed the situation. The first was the arrival in the early part of the century of efficient small electric motors that could run on the growing number of alternating current power systems. It seems curious now that it was easier to build larger motors, and that it took so long for the small ones to be developed. They were vital for a compact domestic refrigerator. Of course, they were turned on and off by the ubiquitous thermostat to maintain the required temperature.

Next was a reliable way of making a system that didn't leak the dangerous refrigerants. The early machines used ammonia, sulfur dioxide, methyl chloride, ethyl chloride or isobutane, none of which were pleasant if they escaped and deaths followed in the severest cases. To really become a commercial proposition a better fluid was required. In 1928, Thomas Midgley Jr., aided by Charles Franklin Kettering, the inventor who had contributed so much to the development of automobiles, developed a new compound—a chlorofluorocarbon—commonly known as Freon.[4] It was harmless to people if it escaped and was efficient in use. Unfortunately, much later it was found to be damaging to the Earth's atmosphere, but at the time it seemed the perfect solution.

To be absolutely sure that no leaks occurred, General Electric and the Westinghouse Company introduced the hermetically sealed compressor in around 1930. This became the standard system; a small electric motor driving a compressor in a sealed system using Freon or an equivalent fluid. With the move of the large electrical manufacturers into the market, the refrigerator was ready to take off in America. In the 1930s they started to sell in their millions and by 1939 more than fifty percent of households had one.[5]

With plenty of space to accommodate them, they could be quite large and later even incorporated ice makers.

In the UK, the situation was very different. The Frigidaire company found moving into Britain was something of a challenge because "ice was regarded as only an inconvenience of wintertime and cold drinks an American mistake".[6] Numbers rose slowly at the rate of some 20,000 per year so that they had reached about 200,000 or one-and-a-half percent of homes by 1939; a remarkable contrast to the situation on the other side of the Atlantic.[7]

Following the war there was still little interest in refrigerators until the 1950s when the numbers sold began to climb, but it still took until 1957 before 10% of homes had one. A hot summer in 1959 helped to give ownership a boost, but it was really during the 1960s when they changed from a minority to majority possession. Around sixty percent of homes possessed them in 1970 and this reached over ninety percent a decade later.

Why the sudden change of attitude? The technology had been stable for some considerable time and refrigerators were deemed perfectly satisfactory in the US long before this. A factor was the increasing numbers of women going out to work. For some 30 years, approximately one-third of women over the age of 15 had been in paid work. By 1961 this had risen to 37% and by 1971 to 43%.[8] At first sight these don't seem large changes, but taking into account that the remainder includes those in education, looking after others, or the 25% who were retired, who are not available for paid work, then the increase is quite significant.[9]

Though the bar on married women working in professions was removed by an Act of Parliament in 1919, it didn't really die out in the depressed conditions between the wars.[10] There was still a strong belief that the woman's place was in the home. It took the labor shortages in World War II to destroy this attitude and the residual rules of some employers. With the post-war baby boom it took until well into the 1950s before the bulk of those mothers felt able to seek paid work.

Women no longer had time for the daily chore of shopping for fresh food, and needed a way of storing it. Or was the availability of the means of storing the fresh food the factor that made it possible for them to seek employment? It would seem that the motivation for purchasing refrigerators was rather different in Britain than in the US, where the love of cold drinks was much greater and large parts of the country were considerably hotter.

In addition, during this time the economy was doing well which meant that there was more money to buy items that had been considered luxuries only a few years before. The increased opportunities meant that there were more jobs for women, particularly in the newer "light" industries such as electronics, and in the ever-increasing numbers of offices.

Gradually, the refrigerator was changing from a luxury to a necessity as the way of life steadily changed. One of the coincident phenomena was the rise of the supermarket. Though there had been experiments with self service for some years, it was only in the 1950s, and particularly the 1960s, that they became ubiquitous. As can be seen in Fig. 17.1, their rise almost exactly matches that of refrigerators.

The less frequent, or maybe once a week, shopping trip was now a reality as more and more people had cars which enabled them to easily carry home the much heavier quantities that this necessitated. It seems impossible to separate all the factors as they are

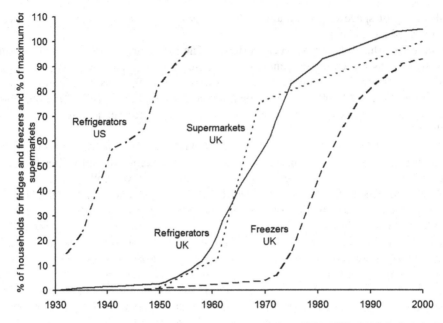

Fig. 17.1 The rise of sales of fridges and freezers in the US and UK, 1930–2000. In Britain it is contrasted with the rise in the number of supermarkets. Source: Author[11]

interconnected and undoubtedly feed off each other. What can be said is that refrigerators were an important factor in changing the way people shopped for food and, hence, in changing their lifestyles in a quite fundamental way.

An American by the name of Clarence Birdseye was born in 1886. He studied biology and then took up work as a naturalist for the US government.[12] From about 1912 to 1915 he worked in Labrador, in Canada, where he learnt the ways of the local Inuit, including how to fish in the very cold climate. He discovered that fish caught at very low temperatures froze almost immediately, but the surprise was that, when thawed months later, they had virtually the same taste and texture as if they were fresh.

After World War I, Birdseye worked on this idea as the results were so much better than any of the frozen food up to that time. He came to realize that the secret lay in the very rapid freezing that occurred in those conditions. Ice crystals grow relatively slowly and quick freezing meant that they remained small and so didn't disrupt the cell structure of the food being frozen. It was a major discovery.

In his spare time he worked on perfecting a freezing method that would produce the same effect. It took him years but eventually he had a satisfactory process which packed dressed fish, meat or vegetables into waxed cardboard cartons which were rapidly frozen under pressure between a pair of cold plates.[13] He took out a whole series of patents to protect his ideas.[14]

This led to the setting up of the General Seafood Corporation, but it wasn't until 1930 that frozen food was finally put on the market. It was sold in a small trial under the trade name Birds Eye Frosted Foods. Birdseye then found that some of the products were melting because they weren't being stored properly and he had to get special frozen food cabinets made for use in the shops. This in turn led to freezer railway wagons to move the products. Birdseye invented the whole concept of the "cold chain".

The first truck with a freezer unit appeared in 1949, which meant that the products could be kept in cold conditions all the way from the freezing plant to the store where the buyer took it out of the freezer and carried it home. It was now generally realized that the food was quite safe as long as it was kept frozen, and only thawed when it was about to be eaten.

By 1950, the whole idea had really taken hold in America where sales reached a billion dollars a year. It was also spreading to other countries. These were not just fresh foods that had been frozen, but increasingly were precooked or prepared dishes. The idea that a ready meal could be frozen after manufacture and then taken home by the consumer and simply heated up, steadily took hold.

It took some time before it was realized that another piece of the jigsaw was still missing. The consumer had no way of keeping the food frozen, so it had to be eaten straight away. As refrigerators became more common and patterns of shopping changed, this was a considerable drawback. Despite this, it wasn't until 1959 that refrigerators began to appear in Britain with small freezer compartments at the top. The rest of the cabinet was just an ordinary fridge.

Of course, the industry was also making deep freezers, which were popular in America, but only sold in small numbers in the UK and were usually large horizontal chests far too big to go in the average British house. Many of them languished in garages. Many owners used them to store half a cow for months, gradually eating it piece by piece. They were also useful for preserving homemade food cooked in large quantities and then stored in smaller packs which could be eaten one at a time as needed.

However, throughout the 1960s the vast majority of frozen food was stored in the small freezing compartments of refrigerators. There was only enough space to keep a tub of ice cream and some of the ubiquitous fish fingers or frozen peas. They weren't powerful enough to freeze the home cooking, and in any case they were soon full.

It wasn't until 1970 that the situation began to change. In that year Malcolm Walker, and another bored young Woolworth's employee, scraped together $90 (£60) and rented a shop. Iceland, the frozen food retailer was born.[15] At first it sold loose frozen food—customers could buy a scoop of fish fingers or peas—but rapidly moved to packages and branched out into its own brand items. Other outlets such as Bejam soon followed, and it didn't take the supermarkets long to join in to reach a promising new market.

Gradually the customers began to see the point of having a freezer with rather more room than the small freezing compartment in their fridge. The numbers of chest freezers rose steadily throughout the 1970s, reaching some fifteen percent of homes by the end of the decade.[16] The problem of accommodating these cumbersome objects still hadn't gone away.

Though a solution had been around for a while, it took the manufacturers a couple of years to develop it into an attractive proposition. Why not stack a fridge on top of a freezer, but sell it as a single unit? The fridge-freezer was born. Though it was far taller than the normal working-top height that had become the norm for fridges, it only had the same footprint, and so it was quite possible to find that much space in the corner of the kitchen. Within seven or eight years the numbers of fridge-freezers in homes exceeded those of chest freezers and they continued to expand rapidly. The answer had been found for British, and generally European, conditions.

Really there is little difference in principle between a fridge and a freezer. The latter only requires better insulation and a slightly more powerful refrigeration unit to enable it to reach and maintain around 0 °F (−18 °C) rather than the less than 40 °F (4 °C) of the refrigerator. There was no fundamental change so it was a fairly easy step for the manufacturers to extend their range.

The growth in the use of freezers was tied to several trends at the time. There was a rise in foreign holidays that brought a greater desire to try different foods, but this was coupled with a demand for convenience food that would cut the work in the kitchen. The answer lay in an ever-increasing range of ready-made frozen foods. These ranged from lasagne to chicken korma to gateaux.[17]

There were some strange results. One of the advantages of frozen food is the availability of vegetables and other items throughout the year and not just when they are in season. It would be expected that this meant consumers began to eat more vegetables, but in fact the opposite happened. Sometimes the offerings were merely a matter of convenience, for example with frozen French fries that only needed heating in the oven. Despite this, the advent of a wide range of frozen food made diets far more varied.

One of the side effects was that in a world that began to worry about preservatives and the amount of E numbers added to food, the fact the frozen meals didn't require any was a considerable bonus. Many suppliers went to great lengths to ensure that crops were really fresh and immediately frozen so that when thawed at home the food seemed as fresh as when it left the field. This was in contrast with so-called fresh food that often had taken many days in the supply chain from source to kitchen.

The range of preprepared frozen meals steadily increased into the 1980s, but then came a change of tack with the introduction of cook-chill technology. This began for commercial establishments, particularly smaller restaurants and pubs, but extended to the home. The principle was to prepare the food under strict clean conditions and then cook it. The next stage was to rapidly cool it using a blast chiller. The meals could then be stored in a refrigerator for five days before being heated up to eat.

The market for these preprepared foods grew enormously in the 1990s and soon they overtook frozen meals.[18] The perception was that they were better value and there was a greater range of recipes. One of the advantages was that it was possible to supply dishes that were too complex for the average home cook to make themselves. The effect of all this was to greatly reduce the average preparation time for home-cooked meals. This was obviously desirable with less time available, and with a generally higher standard of living the money was available to buy these products.

The idea of cooling using the forced evaporation of liquids, coupled with electric motors to provide the power, produced a whole industry. To begin with these devices changed the food supply chains, enabling food to be moved much greater distances from source to consumer. Later, they were to find their place in virtually every kitchen as a reliable, attention-free appliance, which changed the way the people lived, and particularly what they ate.

NOTES

1. *Manchester Guardian*, September 5, 1923.
2. Linde, A.G. 125 years of Linde: A chronicle, available at: http://resources.linde.com/international/web/linde/like35lindecom.nsf/repositorybyalias/pdf_ch_chronicle/$file/chronicle_e[1].pdf.
3. Nagengast, B. Electric refrigerators' vital contribution to households, 100 years of refrigeration. *A Supplement to the ASHRAE Journal*, November 2004.
4. Krasner-Khait, B. The impact of refrigeration. *History Magazine*, Feb/Mar, available at: http://www.history-magazine.com/refrig.html.
5. Bowden, S. and Offer, A. Household appliances and the use of time. *Economic History Review*, XLVII, 4(1994), 725–748.
6. Hardyment, C. *From Mangle to Microwave*, p. 142.
7. Byers, A. *Willing Servants*, p. 57.
8. Calculated from: B.R. Mitchell British Historical Statistics, Labour Force.
9. ONS. Mid-1971 to mid-2010 population estimates: Quinary age groups for constituent counties in the United Kingdom; estimated resident population.
10. Sex Disqualification (Removal) Act 1919.
11. These curves are only approximations as there are no hard figures and they are based on estimates. Some adjustments have been necessary to reconcile the discrepancies between the various sources. The sources for fridges and freezers are: Pugh., *We Danced All Night,* p. 180; PEP Report on the Market for Household Appliances, p. 69; Bowden and Offer, Household appliances and the use of time, *Economic History Review*, XLVII, 4(1994), 725–748; Beynon, H., Cam, S., Fairbrother, P. and Nichols, T. The rise and transformation of the UK domestic appliances industry. Working Paper 42, Cardiff University School of Social Sciences; Boardman, B. et al., Domestic equipment and carbon dioxide emission: Second year report 1995, Oxford University Environmental Change Unit; ONS, General Household Survey—Consumer Durables; Numbers of supermarkets: Universities of Exeter and Surrey, The rise of the supermarket in Britain, available at: http://business-school.exeter.ac.uk/research/consumer_landscapes/shopping/rise.html; Finlay, L., Grocery shopping in the UK: A study of consumers, available at: http://edissertations.nottingham.ac.uk/892/1/07mbalixlcf.pdf.
12. *Encyclopaedia of World Biography* 2004, Clarence Birdseye, available at: http://www.encyclopedia.com/topic/Clarence_Birdseye.aspx; Wikipedia, Clarence Birdseye, available at: https://en.wikipedia.org/wiki/Clarence_Birdseye
13. http://web.mit.edu/, Inventor of the week archive, Clarence Birdseye (1886–1956), available at: http://web.mit.edu/invent/iow/birdseye.html.
14. US patents 1,511,824, 1,608,832, 1,773,079, 1,773,081 to name only a few.
15. Iceland. The Iceland story, available from: http://about.iceland.co.uk/about-iceland/the-iceland-story/.
16. Boardman, B. et al. Domestic equipment and carbon dioxide emission: Second year report 1995. Energy and Environment Programme, Environmental Change Unit, University of Oxford.
17. Food Standards Agency, Eat well, be well—1970s, available at: http://www.eatwellscotland.org/healthydiet/seasonsandcelebrations/howweusedtoeat/1970s/index.html.
18. Food Standards agency, Eat well, be well—1990s, available at: http://www.eatwellscotland.org/healthydiet/seasonsandcelebrations/howweusedtoeat/1990s/index.html.

18

Banishing Washday: Home Laundry

The washing machine had more to do with women's liberation than the birth control pill or their right to work.

L'Ossovatore Romano, the newspaper of the Vatican

Mrs. Blackstone, wife of American merchant William, had a surprise birthday present in 1874—a washing machine.[1] It consisted of a wooden tub, in which was a flat piece of wood with six small pegs. The dirty clothes were hung on the pegs and the board manually swished to and fro in hot soapy water. She was very pleased with it, as it decreased the hard work of clothes washing, one of the worst tasks for the housewife. Though crude, the device was sufficiently attractive for the neighbors to want one too, and led to Blackstone making them commercially.

This was by no means the first attempt at producing a machine to reduce the burden of washday. There had been numerous attempts, dating back to the eighteenth century, to cut the hard work of rubbing the clothes to try to get the dirt out. Along the way it was discovered that rubbing wasn't necessary if the water could be made to slosh through the fabric. This worked much better with the use of hot water and soap (Fig. 18.1).

Mrs. Beeton, in her nineteenth-century *Book of Household Management*, describes a wash-house with hot and cold taps over the tubs for the washing.[2] In practice, few had that luxury and large quantities of water had to be moved in and out of the wash tubs. Then there was the problem of producing the hot water, and often this was in a copper which could be heated with a coal fire or later with gas or even electricity.

The problem with the manual machines was that they didn't reduce the work very much and so were not very popular. Most people just used a wash boiler, like a Burco, and did the rest by hand.[3] What was needed was something to power the machine; the only really practical form for the home was electricity, and this didn't appear in houses to a significant degree until well into the twentieth century.

It is thus unsurprising that patents for electric washing machines appear thick and fast around the middle of the first decade of the century. As connecting an electric motor to a manual washing machine is fairly obvious it is unlikely that any of these are really valid. The one most quoted, from Alva J. Fisher, doesn't even claim the use of an electric motor

© Springer International Publishing AG 2018
J.B. Williams, *The Electric Century*, Springer Praxis Books, DOI 10.1007/978-3-319-51155-9_18

Fig. 18.1 Traditional washday equipment of tubs and washboards for rubbing clothes.
Source: http://www.oldandinteresting.com/washboards-history.aspx

as an invention.[4] What is important is that commercially-available machines appeared around 1906 or 1907. Probably the 1900 company using T J. Winans' design just beat the Hurley Machine Company of Chicago's Thor electric washer to the market.

The exposed nature of these early machines was enough to frighten the average laundrymaid. They featured a paddle to agitate the water in the tub, and usually had a powered mangle mounted above to squeeze the water out after washing. Looking at them, it is hardly surprising that they didn't really catch on. That, coupled with a relatively high price and the small number of homes that had an electric supply, meant that their time hadn't come.

Within the next few years a vast number of companies piled into the market, but perhaps the most notable was the Maytag company. It had started out as a farm machinery manufacturer and moved into washing machines "in the slow time". After World War I it became one of the main manufacturers, and introduced aluminum tubs instead of wood as a step along the way to a modern machine.

It is noticeable that most of the development happened in the US. There are many reasons for this, but a major factor was the more rapid electrification of homes. This isn't the whole story as when the electric supply to British homes began to catch up in the late 1930s the sales of washing machines in America were some two million a year compared with 60,000 in the UK.[5] Perhaps more important were the continued availability of servants in Britain and the pure affordability or otherwise of the machines.

Most of the early machines used the conventional wringer or mangle to remove the bulk of the water from the washing. The only advance was that it was now powered which, while it reduced the work, also made it a lot more dangerous. During the 1920s another device for removing the water began to appear, based on a completely different principle—the spin dryer. A perforated drum is spun inside another drum and the water is thrown out of the washing by centrifugal force. It is quick, and remarkably effective, but at the time few people could afford the high cost of another machine, particularly when they already had a mangle or wringer either separately or as part of their washing machine.

The final step in the washing process was to get the items dry. Here the simple washing line was much the cheapest option, though in wet weather for those without the luxury of a drying room it meant wet washing hanging around the house. To address this problem, yet another machine appeared—the tumble clothes dryer. It is quite simple in principle with a slowly rotating drum through which hot air is blown. Despite being effective, the numbers sold were very small.

The Holy Grail was to bring all these things together in a single machine into which the dirty washing was fed and clean, dry items came out. It didn't look too difficult as all the essential parts were already in existence. A cam timer could be used as a sequencer to turn things on and off at the appropriate points and allow water in and out. The motor could be switched from low speed for the washing to a higher one for the spinning.

The trouble was that was far too expensive for the average customer. Various types of semi-automatic arrangements were tried and had limited success. Probably the best attempt was by Bendix who produced a machine shortly before the war. It had a horizontal drum, instead of vertical, with projections on the inside which worked by lifting the washing out of the hot soapy water and letting it drop back in. After a suitable number of wash and rinse cycles, it rotated a high speed to remove most of the water.

It was the Bendix and similar machines which made possible a solution to the washday problem—the launderette. A number of automatic washers were fitted with coin slots and installed in a small shop. Anyone could come in and load their dirty laundry into the machine and wait while it was washed and spun. Also available were large tumble clothes driers which, for another fee, would dry the load. It was a simple idea and the high usage made the expensive machines economic.

They became quite common in America after the war, but the first one only appeared in Bayswater in London in 1949. They became quite popular, particularly for students or those living alone and by 1970 there were some 7000 in Britain.[6] However, many people regarded them as rather down market and their aspiration was to have a washing machine of their own even if it was less automatic than the launderette machines.

In most American houses, there was plenty of room and the size of the machine was not an issue. What is remarkable is that, despite the depressed conditions in the 1930s, the growth in the ownership of washing machines was uninterrupted (Fig. 18.2). There was a slight slowdown immediately after the stock market crash in 1929, but the upward trend was soon resumed and, if anything, accelerated. By the war more than fifty percent of American homes had a machine.

In Britain, with tiny kitchens common in all those semi-detached houses built in the 1930s, size was critical. The first company to really get it right was the American company Hoover, best known for their vacuum cleaners. In 1948, they built a factory at Pentrebach, Merthyr Tydfil in South Wales to make an entirely new machine, which was small enough to go under the draining board in British kitchens, but large enough for the average wash. It was neat and tidy with no exposed working parts and suitable for use by the housewife.

It was a very simple machine with a paddle in the side of the tub to agitate the water. This produced a cheap machine at a little over $45 (£30), but it did tend to tangle the clothes.[7] With a pump to get the water out, and a hose to fill it from the tap, it could do the weekly wash efficiently. A foldaway manual mangle at the top provided for the removal of the water. As the company's managing director, Charles Colston, said: "It is not an expensive model intended for the few. It has been built with the intention that it shall be for the million."[8]

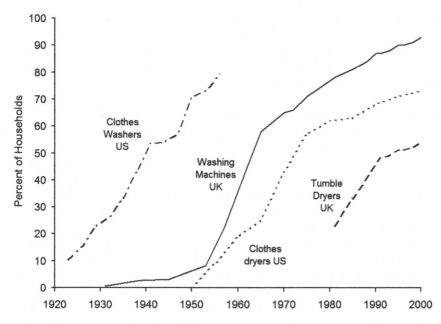

Fig. 18.2 Percentage of homes in the US and UK with washing machines and tumble dryers, 1920–2000. Source: Author.[9]

Though at first a large proportion went for export in the late 1940s, when the country was desperate to earn foreign currency, the way forward was now clear. Very soon other manufacturers were producing something similar though most went for an agitator in the bottom of the tub which rotated backwards and forwards to minimize the tangling. As a result, the British market for washing machines started to take off. The industry had discovered the secret, which was to keep the price and size of the machine down even if this meant that it only did part of the job.

One of the things that made these washing machines practical was the arrival of synthetic detergents that removed dirt much more efficiently than the soap that had previously been used. This had been a particular problem in hard water areas where the soap produced a scum. Though the first-generation detergents appeared in the 1930s it was only after the war with the introduction of improved formulas that they really started to be used in large quantities.[10]

As the 1950s progressed there was a rise in interest in spin dryers as small, stand-alone units appeared, but it was the next step, almost unique to Britain and her dominions, that had a major impact. It was an obvious idea to combine the washer and the spin dryer in a single cabinet—the twin-tub was born. Once again it was Hoover that introduced them in Britain in 1957, but other manufacturers such as Hotpoint and Servis soon followed.[11] The twin-tub was still a quite simple machine and it required some work from the housewife, particularly lifting the laundry from the washtub into the spinner to extract the water. Nevertheless, it was a considerable step up from the mangle, and the machines could still be made economically.

Generally, the machines were quite reliable although the Adamatic, made by Ada of Halifax, had a notorious fault. The way all the machines achieved safety for the spinner was that the motor was only powered when the lid was closed. When the lid was lifted the power was cut off and a brake applied. In the Adamatic, the arm operating the power switch often became loose, leading to the brake coming on while the motor was still powered. The result was a kitchen full of acrid smoke due to a burnt-out motor.

It was at this point that another player emerged. His name was John Bloom, the son of an East End tailor, and he turned out to have a flair for selling things to housewives. He began selling Dutch-made washing machines door to door, moved on to having them made in Britain, but really got going when he managed to get control of the semi-defunct Rolls Razor company and used that as a brand name for his machines. He placed display adverts in newspapers, with a coupon to send back to get free information which would soon be followed up by a salesman.

The secret was that, by selling direct and on hire purchase, he could price the machines at something like half that the mainstream manufacturers were charging, who were using a complicated distribution system ending in local shops with each level taking their cut. Within a year or so, he had grabbed ten percent of the market and the competition was trying to find ways of stopping him. Eventually, they simply reduced their prices and Bloom's castle came crashing down. He had been paying very large dividends (particularly to himself) and the company ran out of money once the sales dipped and the financial backer took fright.[12] Despite his spectacular failure, he had lowered the price of washing machines and, hence, increased the number of people who could afford them, particularly on hire purchase.

Despite the success of twin tubs, what the housewife really wanted was an automatic machine that meant she could push in the washing and forget it. With more women going out to work, both the money and the requirement were there. The Bendix had shown the way but it had one major drawback—it needed to be bolted to a solid floor because of the vibration when it started spinning the clothes to get them dry.

The Bendix had used a horizontal drum with the load inserted through a door at the front rather than at the top. In America, this top-loading arrangement was more popular as there was plenty of space for the machines and the manufacturers retained that. In Europe, with less space, the machines usually had to fit under a worktop or counter, so the front-loading arrangement was more suitable. However, in each case there was a need to get away having to bolt the machine down.

The solution was to suspend the drum from springs and use dampers in a similar manner to a car suspension. Here the idea was to ensure that the vibration was not communicated to the outside, rather than preventing rough roads being felt inside. The result was that the machine could be stood on the floor without needing to be bolted down. Only if the load was badly unbalanced would it move about but this could be controlled by stopping the spin cycle and tumbling the washing again to try to get a better balance. The way was now open for domestic automatic machines.

There was still a problem with the spinning—obtaining a sufficient speed. A front-loading machine with a typical sized drum really needed to run in excess of 1000 rpm to give a satisfactory spin dry. (The twin tub spinners ran at around 3000 rpm but the drum was smaller.) One way of achieving the slow speed for wash (around 50 rpm) and then the high spin speed was to use two motors, but this was expensive and only used by the top-end machines, such as the German Miele.

A cheaper solution was to use a multi-pole AC motor. If the motor had 12 poles it would run at 500 rpm which could be geared down with a belt and pulleys to drive the drum at the wash speed. By switching to only using two poles, the motor speed would be around 3000 rpm, giving a spin speed of 300 rpm. This was what the cheap machines did. It might have been adequate in the warm climate of Italy where many of them were made, but in Britain, and most of Northern Europe, the washing came out far too wet. With a bit more money an 18-pole motor could be used give a spin speed of 450 rpm, which still wasn't really adequate.

In the late 1960s, electronics came to the rescue. By using a DC motor, connected to a thyristor to control the voltage, high spin speeds could be achieved economically.[13] It was also easy to add other features, such as different speeds, and ramping the speed gently through the point where the tumbled washing starts to stick to the sides of the drum. At last, a general-purpose domestic machine could be made at a price that people could afford and the sales took off.

Though some of the early machines were not as reliable as was desired, over the years components such as water level switches improved, leading to fewer floods in the kitchen. There was, however, one part that defied improvement without the cost running away and that was the timer. It was a complex electromechanical device that controlled the whole sequence from the reverses of the drum during washing to the rinses and spins.

Towards the end of the century the arrival of suitable low-cost microprocessors allowed these small "computers" to take over the duty. They allowed more flexibility but, not being full of cams, springs and bits of bent metal, the reliability was improved. The result was a machine suitable for every kitchen or utility room that virtually all families could afford with the result that nearly all had one.

The split remained; in Europe the standard machine had front loading and a horizontal drum so that it would fit under a counter or work surface. In America, with typically more space in houses, and the machines often installed in basements, there wasn't this requirement and top-loading machines with a vertical drum remained popular, despite the fact that they used considerably more water.

While these machines made a good job of washing the clothes and removing most of the water, there was still the issue of getting the laundry dry. Many people preferred to hang it up, particularly outside, so that it dried naturally. Washing allowed to flap in the wind came out less stiff than if dried statically. For many people this was a reasonable compromise, particularly as the ironing had to be done by hand.

So what of the tumble clothes dryer that had been invented many years before? By comparison it was a relatively simple machine, only requiring a drum to turn, ideally first one way and then the other, while hot air was blown through it. Curiously, they have never achieved their full potential of one in every home like washing machines. For some, the problem is finding space for yet another machine, but for many it is the pure cost of running it when simply hanging up the washing allows it to dry without expense. The result is that, by the end of the century, still only about half of British homes had one, and even when they did it was not used all the time. Even in America it had not reached three-quarters of homes.

Why not combine there two machines into one? They share many features such as the drum and timing systems. At first sight this appears a straightforward solution. There are no serious problems in producing such a machine. The reasons why they are not popular are all to do with the acceptability to consumers. They are more expensive than a washing machine, but still suffer from the same issue as with the tumble dryer—it is much cheaper to simply hang out the washing.

The only technical matter is that wet washing takes up less space than dry. The result is that a bigger drum is required for drying than for washing. With the machine needing—certainly in Europe—to be a standard size this means that it is going to wash a smaller load than a normal washing machine. These factors have conspired against the composite machine, and certainly in Britain they have never been popular.

Have washing machines really lived up to the hype of liberating women? After all, they don't dry or iron it, and these tasks still have to be done by hand. However, you only have to ask someone whose machine has broken down. They want it fixed immediately as the pure effort of washing by hand is considered too great a task for the weekly wash, particularly now that clothes are changed much more often. In that sense, it has been liberating, even though women are still involved in most of the time spent on laundry work.[14]

NOTES

1. AMDEA, The Association of Manufacturers of Domestic Appliances, Washing machines, available at: http://www.amdea.org.uk/industry-information/our-products/washing/.
2. *Mrs. Beeton's Book of Household Management*, available at: http://www.mrsbeeton.com/41-chapter41.html#2373.
3. Hardyment, C. *From Mangle to Microwave*, p. 62.
4. US patent no 966677; The whole subject of who invented the electric washing machine is discussed in L. Maxwell, Who invented the electric washing machine? An example of how patents are misused by historians, available at: http://www.oldewash.com/articles/Electric_Washer.pdf.
5. Hardyment, C. *From Mangle to Microwave*, p. 64.
6. Harrison, B. *Seeking a Role: The United Kingdom 1951–1970*, p. 163.
7. Hoover advertisement from 1950 available at: http://www.historyworld.co.uk/advert.php?id=303&offset=0&sort=1&l1=Household&l2=Appliances+%28large%29.
8. Kynaston, D. *Smoke in the Valley: Austerity Britain 1948–51*, p. 38.
9. These curves are largely based on estimates. Some adjustments have been necessary to reconcile the discrepancies between the various sources. The sources for washing machines are: M. Pugh, *We Danced All Night*, p. 180; PEP Report on the Market for Household Appliances, p. 69; Science Museum, Making of the modern world, available at: http://www.makingthemodernworld.org.uk/stories/the_rise_of_consumerism/02.ST.03/?scene=3; D. Bowden and A. Offer, Household appliances and the use of time, *Economic History Review*, XLVII, 4(1994), 725–748; H. Beynon et al., The rise and transformation of the UK domestic appliances industry; B. Boardman et al., Domestic equipment and carbon dioxide emission: Second year report 1995; General Household Survey – Consumer Durables; Tumble dryers from General Household Survey, no figures were found before 1981; US washing machine figures also from Household appliances and the use of time, but adjusted to be for all households, dryer figures from: R. McGrath, The pace of technology is speeding up, *Harvard Business Review*, November 25, 2013, available at: https://hbr.org/2013/11/the-pace-of-technology-adoption-is-speeding-up/.
10. Cope, H, The history of detergents, available at: http://www.chem.shef.ac.uk/chm131-2001/cha01hc/dhistory.html.
11. Byers, A. *The Willing Servants*, p. 68.
12. *Time*. Bloom at the top, October 13, 1961; Trouble in never never land, July 24, 1964.
13. In more recent years, other schemes such as AC motors driven from inverters have been used, but the principle of electronic control of the motors to achieve the large change in speed is maintained.
14. Bittman, M., Rice, J.M. and Wajcman, J. Appliances and their impact: The ownership of domestic technology and time spent on household work. *The British Journal of Sociology*, 55:3, 2004.

19

Going Up… or Down: Elevators and Escalators

Vertical transportation is now generally recognized as being of as great importance as horizontal transportation… [by]great users such as railway companies, department stores, hotels, office buildings, and other edifices.

Henry C. Walker, Chairman, Waygood-Otis, 1929

In 1880, the city fathers of Mannheim, a small city at the confluence of the Rhine and Neckar rivers in Germany, decided to hold a Trade and Agricultural Exhibition. Looking around for an attraction to pull in the crowds they approached Werner von Siemens. He had had considerable success at the Berlin exhibition the previous year with a small electric train that ran around a 850 ft (270 m) circuit. This time, he was asked to build an electric elevator or lift which would go up a temporary tower and give views over the city from the top.

The short time available and the fact that no-one had built one before didn't concern von Siemens, and he accepted the commission. Though delivered part way through the exhibition it was a great success and, despite charging 5 cents (20 pfennigs) a ride, in two and a half months it carried 8000 people.[1] It used an ingenious system where the motor and gearing were mounted under the platform which climbed up a rack mounted on the tower (Fig. 19.1).

With many other interests, Siemens seems to have done little further in the field. He appears to have only built one more elevator and that was a decade later at the Mönchsberg near Salzburg which also used the rack and pinion principle for its 190-ft (60 m) rise[2] which is similar to the rack and pinion system used on mountain railways. Siemens was ahead of his time and electric elevators were not seriously to appear for a number of years.

Siemens' demonstration was, however, part of a tradition of using elevators as an attraction at such events. Elisha Otis, another American farm boy turned inventor, had made his name by demonstrating his safety "brake" system at the New York Crystal Palace exhibition in 1854. This was an adaptation of the simple hoist that had been used for many years for raising loads, but Otis' important addition was a locking system that held the platform if the rope was cut. It was a vital step in making these devices

© Springer International Publishing AG 2018

J.B. Williams, *The Electric Century*, Springer Praxis Books, DOI 10.1007/978-3-319-51155-9_19

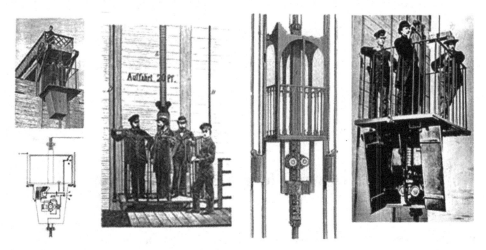

Fig. 19.1 The first electric elevator created by Werner von Siemens for the Mannheim exhibition. Source: http://www.theelevatormuseum.org/elec1.php

a practical proposition for passengers as well as goods. Quite soon his company, and others, were producing elevators powered by steam engines.

A better motive power was provided by hydraulics. This could use the local water supply if it had enough pressure, or a pump if it didn't. The elevator was raised by the water entering a cylinder and forcing out a piston, which could be under the cabin pushing it up in the simplest form, though this needed a pit as deep as the distance the cabin needed to travel. To avoid this, ropes and pulleys (sheaves, as they were known) could be used to pull it up. This gave the flexibility for the hydraulic cylinder to be mounted in the shaft alongside the cabin, or horizontally at the base.

The Eiffel Tower was built in 1889 as the main attraction of the Universal Exposition in Paris to celebrate the centenary of the French Revolution. It was essential that it had elevators as it was 1000 ft high (300 m), and intended for people to use as an observation tower. In fact, there were three sets of elevators by different manufacturers, all of them hydraulic. Though an electric scheme was proposed it was rejected, probably because it was felt that the technology wasn't sufficiently mature.[3]

The Eiffel Tower set a fashion for tall structures. In England, there were four schemes, three of which were near copies in various sizes. The first was Blackpool Tower in 1894 which had hydraulic elevators.[4] The ill-fated Wembley Tower was only part built due to unstable ground.[5] By the time the third one at New Brighton on the Wirral arrived, the elevators were electric, as were those intended for the uncompleted tower at Morecambe.[6] The great heyday of hydraulics was from around 1880 until around 1895, after which electric elevators became a practical proposition.

In 1873, Henry Hankey, described as a merchant and City banker but in reality a piratical developer, started to build a block of flats called Queen Anne's Mansions in London.[7] They lay between St. James' Park and the station of that name, and to maximize the potential of the site his plan was to build upwards. He ignored the regulations preventing buildings over 100 ft (30 m) in height until he reached that level, and then applied to get retrospective permission. This first section topped out at 116 ft and ten storeys, but over the next 15 or so years the successive stages went to 130, 141 and finally 160 ft. Hankey had realized that by installing elevators the upper floors could easily be reached and would still command high rents.

People were horrified by these out-of-scale buildings, and it is claimed that Queen Victoria complained that they blocked the view of the Houses of Parliament from Buckingham Palace. The result was the London Building Acts restraining heights to 90 ft in 1890 and then 80 ft in 1894. In one fell swoop, the possibility of building skyscrapers, as was becoming popular in America, was removed.

At first sight, this would appear a blow to the development of elevators in London, but the mansion blocks were popular for people with money and, like in cities around the world, many were built in the later part of the nineteenth century. Elevators were an essential part of the building, though at first they used hydraulics, usually powered from water tanks in the roof like at Queen Anne's Mansions. However, by the 1890s electricity undertakings were starting up in these cities and it soon became the fashion to have electric lights in these blocks. Once there was an electricity supply it made sense to use that to also power the elevators. From 1892, that became a feature when advertising vacant mansion flats.[8] Electricity was the coming thing and it had more kudos than the by now old-fashioned hydraulics.

One of the reasons that elevators were practical in these blocks was that attendants could be employed to run them. A hydraulic elevator was controlled from the cabin by a rope passing through it. The attendant pulled down on the rope which opened the water valve, allowing water into the cylinder and causing the cabin to rise. When it reached the required floor, the attendant let go of the rope which activated the brakes. To go down, the rope was pulled up to allow water out of the cylinder, allowing it to descend. The problem was that the rope was fixed to the building and not the cabin so it had to be paid out as it traveled, and this required some skill.

At first the electric elevators used a similar system, but now the starting and stopping of the motor was controlled.[9] However, it wasn't difficult to bring wires to a controller in the cabin which could be turned one way for up and the other for down. Also the doors, both in the cabin and on the landings, could be interlocked so that the elevator couldn't be moved until all the doors were properly shut. This led to a considerable increase in safety.

Another advantage of electric systems was that more sophisticated control could be introduced. The push button system was introduced in the late 1890s.[10] On each landing there was a button to call the elevator. When it arrived, the user opened both the landing and cabin doors, carefully closing them behind them. Inside there was a set of buttons, one for each floor. Pressing the required button would mean that the elevator would travel to that floor where the reverse process was required with the doors.

This system only worked reasonably well in places without much traffic because it suffered from a couple of drawbacks. Firstly, the user had to remember to close the doors after they left because the elevator couldn't work at all if the landing doors were not shut and locked. Secondly, when it was in use no-one else could even call it because it would only respond to a call when the first user had finished, even if this meant that it went right past where the second user was waiting.

It is normally accepted that the first department store to install an elevator was New York's Haughwout Building on Broadway in 1857. This was an Otis machine powered by a steam engine in the basement. It set a trend for such installations in America where tall shop buildings were common. There is a tendency for shops in a similar trade to huddle together as this is good for trade—in modern parlance, it increases the footfall. This phenomenon went further in the US, causing land values in those areas to increase and the stores to go higher and higher, bringing the need for vertical transportation.

In Britain, the situation was different. Department stores grew piecemeal mostly out of small drapery stores and that, coupled with planning restraints, meant that they only traded from ground and sometimes first floors.[11] Thus there was not the same requirement for the customers to be transported to upper floors and elevators came much later, usually when a major rebuilding was necessary. Jenners in Edinburgh burnt down in 1892 and was rebuilt, opening in 1895 with electric light and hydraulic elevators.[12]

Harrods built higher, but had elevators to flats above the store which only occupied the lower floors. Only later did it gradually incorporate the upper floors of the shop itself. When the American Gordon Selfridge opened his store in Oxford Street in 1909 he followed the normal practice from his home country to look after his customers and installed "smooth running electric elevators to whish them comfortably from floor to floor obviating all fatigue and saving time".[13] He set a standard that others had to match.

There were those who thought that there was a better solution to move streams of people between levels. One such was the "Paternoster", called after the first two words of the Lord's Prayer in Latin, either because it was like the rosary beads or possibly because you needed to say a prayer before entering one! It consisted of a series of open-fronted cabins on an endless chain which traveled slowly but never stopped. The user needed to step into the moving cabin as it passed and step out again at the destination floor.

They are often thought to have been invented by J.E. Halls of London in 1884, who called it the cyclic elevator, but they may well have existed earlier than this.[14] They were never popular in Britain, but there are still a few in existence such as in the Attenborough Tower of the University of Leicester. They are more common on the European continent, particularly in Germany. They are only really satisfactory in places with agile young people, such as universities.

Another approach was to have a moving ramp or staircase. There were numerous attempts at this in the nineteenth century, but it only came to fruition right at the end. The first ones were moving ramps. In 1892, the American Jesse W. Reno patented an endless conveyor though it wasn't actually produced until about 1897 when one was installed at an amusement park at Coney Island, New York.[15] It was only a belt with cast-iron slats for grip.

The French company, Piat, produced one in 1898 which was installed in Harrods in London to speed the customers to the first floor. Again, it was a "stepless" belt made of leather sections linked together. At the top, there were reported to be attendants with smelling salts and brandy to revive anyone overcome by the experience![16]

By 1900, the American Charles Seeberger, working with Otis, had produced a stepped device. It was exhibited at the 1900 Paris Exposition Universelle in France. Though better than the inclined belts, it had flat steps, and at the top there was a horizontal section and the belt disappeared under a triangular divider which forced the passenger to step off to the side.

Of course, an elevator can take the passenger down as well as up. The first London underground railways were not far below the surface, having been made by the "cut and cover" method. They could reasonably easily be reached by stairs. However, when the City and South London Railway (C&SLR) was considering their line in the 1880s the decision was made to go deeper, partly to avoid the disruption but also to avoid underpinning buildings, diverting sewers and pipes, and interfering with cellars of houses.[17]

With the platforms some 70 ft down, it was an essential part of their plan to use elevators to enable the passengers to reach the trains. The original intention was to haul the trains with cables but this was changed at a late stage to electric traction. However, when the line opened in 1890, the elevators remained as hydraulic units built by the company of Lord Armstrong of hydraulic cranes and guns fame. To avoid having to dig even deeper the cylinders were mounted vertically in the shafts alongside the cabins and operated by ropes and sheaves.

When the Central London Railway (the central part of the modern Central Line) was being built in the late 1890s it too needed elevators, 48 of them.[18] The company turned to the American Frank Sprague, who had taken up making them after his work on streetcars and trams outlined in Chap. 4. Together with his partner, Charles R. Pratt, he produced an electric arrangement which was remarkably like the horizontal cylinder hydraulic system. Instead of the cylinder they used a large screw and nut driven by an electric motor to move the sheaves that controlled the ropes which lifted the cabin. Though the system in some ways was rather crude, Sprague did make considerable improvements in the control system. Despite a clever circulating ball system for the nut, this was the weak point and the system didn't go a lot further.

As the new century opened the C&SLR extended their line to the north and the south. They decided to convert to electrical elevators which gave considerable cost advantages, though they did have some trouble getting the method of control satisfactory. They ended up with a set of switches in the cabin rather than the rather crude rope control of the hydraulic system.[19]

These were made by the same company that had constructed those for the New Brighton Tower, Easton, Anderson and Goolden of Erith. They used a better arrangement to drive them. Many early electric elevators had gone back to the drum hoist as used in coal mines. Here, the cabin is hauled up by a rope which winds around a drum. These used a sheave, driven by a geared motor, with the rope passing around it on its way to the counterweight. The arrangement depended on friction but had the great advantage that there was no theoretical limit to how far the cabin could travel. The other systems were either limited by the length of the screws in the Sprague system or the size of the drum.

In 1906/1907, the set of lines opened that were controlled by Charles Yerkes' London Electric Railways company used 140 Otis electric elevators. These had automatic acceleration, deceleration and positioning on the landings. The entrance gates were worked by hand but the exit ones, on the other side, used pneumatics.[20] Now that reasonably satisfactory elevator systems were available, underground railways, Metro or subway systems started to appear in other cities such as New York, Paris and Boston.

The golden year for elevators on the London Underground was 1907, when their numbers reached a maximum of 249,[21] and the vast majority of these were electric. Hydraulics were only used on the original part of the C&SLR and a few other places where it was convenient for some reason. Though there had been considerable argument, the facts were that the electrics were cheaper, both to buy and to run.[22] Once the control issues had been mastered they were taking over almost completely.

However, despite the fact that some of them could take 50 people at a time, they still couldn't move people quickly enough. For some strange reason, passengers will wait much longer for trains or buses than for elevators before getting frustrated. The worst place for problems in London was at Earls Court where large numbers of people would arrive on what is now the Piccadilly line to go to the exhibitions. The elevators couldn't cope and so in 1911 the London Electric Railway company installed a pair of escalators which they bought from Otis in America.[23]

They were a great success with some 550,000 people using them in the first month, averaging 18,000 a day with a maximum of 24,500 on one particular day. The escalators were stepped and were still of the Seeburger type with the flat steps and there were the angled barriers at the exits so that the users stepped off sideways (Fig. 19.2). Later, in 1924, the Underground railways were to convert to ones using ribbed steps and combs at the ends which were safe for the users to step off forward, and is the form still used.[24]

From here on, most underground railways installed escalators in all their new stations, and in older ones when they were upgraded. Their carrying capacity was much greater and they didn't require an attendant. The result was that the number of escalators increased as the underground systems expanded, reaching 303 by the end of the century. During the same time the number of elevators fell to only 64.[25]

This trend was eventually picked up by the department stores in the 1930s. D.H. Evans installed escalators in their refurbished store in 1935 while Boots started using them in 1937.[26] However, the great move into them was after World War II when they became regarded as essential. By the end of the century no large store was without them, and they became an important feature of the new type of Main Street, the shopping mall. It meant that upper floors had the same value as the ground floor as shoppers would quite happily ride to the appropriate level.

Fig. 19.2 The first escalator at Earls Court station. The sideways exit can be seen in the inset
Source: http://www.theelevatormuseum.org/esc4.php

As buildings became taller than around six storeys in the late nineteenth and early twentieth centuries, the focal point of the building changed from the main staircase to the elevator core. As the number of floors increased so did the demands on the elevators. When the 750 ft (240 m) Woolworth Building in New York was completed in 1913 it required 34 elevators to service its 57 storeys.[27] The Willis Tower in Chicago, completed in 1974, and holding the record for the highest building until almost the end of the century, needed 104 to service its 108 floors.[28]

Not only did the numbers increase, but to provide a reasonable service, the speed had to as well. This required a technical advance. The geared motor system meant there were limits on how fast the cabin could be moved. However, by doing away with the gearing and using special motors that could also operate at lower speeds, the elevators could be moved much faster. Whereas the geared units could reach up to around 200 ft/min (1 m/s), the gearless ones have achieved 1800 ft/min (9 m/s) or more.[29]

With the increase in speeds came a more severe requirement in accelerating, and particularly decelerating, the cabin to ensure that it positioned itself correctly at the landing. This required more sophisticated electrical controls. Led by Otis and Westinghouse, these were introduced in the 1920s and 1930s.[30] Also needed was better sequencing of the operation so that the attendant wasn't needed. The simple "single user" control system was just about adequate for infrequent use, but was totally unsatisfactory in heavily-used commercial buildings.

The next step was to introduce relay control systems that could make better decisions on where to stop. Instead of a single call button the user pressed one to indicate whether they wanted to go up or down. This allowed the elevator that was already in motion to decide whether to answer the call if it was going in the desired direction and not if it wasn't. When it had answered all the requests in that direction it would either stop and remain idle or reverse and service any requests in the opposite direction.

Though these improvements had been developed well before the war, it was only in the 1950s and 1960s that they came into general use and the attendants steadily disappeared. However, if there was more than one elevator servicing the same floors, then the control became even more complicated as there was the additional problem of deciding which cabin should answer the call. Control systems moved from relays, to electronics and finally to microcomputers to implement ever cleverer strategies.

While buildings had been getting higher and higher in the US, particularly in New York, planning restraints had prevented this in London. Only in 1958 were they relaxed.[31] This led to a number of tall structures, such as the telecommunications tower (BT Tower) and Centerpoint on Tottenham Court Road in 1964 and then the NatWest Tower in 1980. As the century drew to its close, tall commercial buildings became the fashion around the City of London, like most large cities.

Often the elevators became a feature instead of being tucked away. They could appear on the outside, as in the Lloyds Building in the City of London, or internally at the side of large atriums. What separated these from what had gone before was that they had windows so that the users could see out as they traveled up and down. In fact, many of them were almost completely made of glass and became an architectural feature rather than a pure necessity. Also in the Lloyds Building the escalators became a feature of the central atrium. Vertical transportation had finally arrived.

After World War II, planners and architects looking for a solution to urban slums, felt the solution was to go upwards and put people in tall blocks of apartments. This had the advantages of accommodating a similar number of people, and preventing urban sprawl. New building techniques, but particularly the now fully-automatic elevator, made this a practical proposition. In the 1950s and 1960s tower blocks began to spring up in all the major cities, and by the early 1980s there were some 440,000 homes in them just in the UK.

The net result was that by the end of the century there were some quarter of a million elevators in the UK, the vast majority in various commercial buildings and relatively few in residential. The situation was quite different in many continental countries where living in blocks of apartments was common. Germany, France, Greece and particularly Italy and Spain all had more elevators in residential properties than the total for the UK. Spain had three times the amount just in residential buildings. There were even 75,000 escalators spread across the countries of the European Union.[32]

In the US, there are some 900,000 elevators making some 18 billion passenger trips per year.[33] In addition, there are 35,000 escalators which transport some 18 billion passengers per year. The Otis Company alone has installed 1.2 million units, and they calculate that elevators move the equivalent of the world's population every 72 h.[34]

By the end of the century, a hotel that didn't have elevators would lose guests, shopping centers and department stores without escalators (and a few elevators for the infirm) wouldn't attract shoppers. Modern cities wouldn't be defined by their skyscrapers without the elevators to make them practicable, and sightseeing platforms on tall towers wouldn't be a draw if it meant a lung-busting climb up hundreds of stairs. Elevators and escalators have become so ubiquitous and essential that the world would be very different without them.

NOTES

1. Kiuntke, F. Siemens history, News archive, First electric elevator, available at: http://www.sie-mens.com/history/en/news/1043_elevator.htm.
2. Mönchsbergaufzug, SalzburgWiki, available at: http://www.salzburg.com/wiki/index.php/M%C3%B6nchsbergaufzug; and Siemens Österreich, 1890–1899, available at: https://www.cee.siemens.com/web/at/de/corporate/portal/SiemensInOesterreich/History/1890-1899/Pages/1890-1899.aspx; these are both in German.
3. Vogel, R.M. Elevator systems of the Eiffel Tower, 1889, available at: http://www.gutenberg.org/files/32282/32282-h/32282-h.htm.
4. Wikipedia, Blackpool Tower, available at: http://en.wikipedia.org/wiki/Blackpool_Tower.
5. *Skyscraper News*. Wembley Park Tower, available at: http://www.skyscrapernews.com/buildings.php?id=66.
6. Electric lifts for the New Brighton Eiffel Tower, *The Engineer*, 6 January, 1899; Cooper, G. Towering presence for a new funland. *The Visitor*, 10/2/14, available at: http://www.thevisitor.co.uk/news/local/towering-presence-for-a-new-funland-1-1208943.
7. Dennis, R. 'Babylonian Flats' in Victorian and Edwardian London. *The London Journal*, 33:3, 233–247, 2008.
8. An example is in *The Times*, Saturday, November 5, 1892, p. 4. Many more can be seen in those pages over the next 8 years.
9. Ravenshaw, H.W. Electric lifts and cranes. Minutes of the Proceedings of the Institution of Civil Engineers, PART 4, pp. 11–21, January 1897.

10. Stratosch, G.R. *Vertical Transportation*, p. 121.
11. Shepphard, F.H.W (Ed.), Survey of London Volume 41: Brompton, available at: http://www.british-history.ac.uk/report.aspx?compid=50006.
12. Jenners of Edinburgh A short history, available at: http://travel.ciao.co.uk/Jenners_Department_Store__Review_5315516.
13. From a Selfridge Display advert in *The Times*, Thursday, March 18, 1909.
14. Wikipedia. Paterrnoster, available at: http://en.wikipedia.org/wiki/Paternoster; The Elevator Museum, Paternosters continuous vertical passenger conveyors, available at: http://www.theelevatormuseum.org/esc1.php; Dartford Technology, Escalators and lifts, available at: http://www.dartfordarchive.org.uk/technology/engin_hall_lifts.shtml.
15. US patent no 470918.
16. Wikipedia. Escalators, available at: http://en.wikipedia.org/wiki/Escalator.
17. Greathead, J.H. The City and South London Railway. *Proceedings of the Institution of Civil Engineers*, Part 1, pp. 39–73, January 1896.
18. Croome, D.F. and Jackson, A.A. *Rails Through the Clay*, p. 540; some sources say 48 and others 49.
19. MacMahon, P.V. The City and South London: Working results of the three wire system applied to traction etc. *JIEE*, 33, 100–160, 1903–1904.
20. Croome, D.F. and Jackson, A.A. *Rails Through the Clay*, p. 87.
21. Ibid., p. 541.
22. Crews, H.C. Lifts and hoists. JIEE, 37: 180, 245–260, 1906 and the discussion about this paper in the following pages.
23. Badsey-Ellis, A. The first escalators. London Underground Railway Society, *Underground News*, May 2007.
24. Croome, D.F. and Jackson, A.A. *Rails Through the Clay*, p. 542.
25. *The Times*, Wednesday, June 16, 1999, p. 4.
26. *The Times*, Monday, January 21, 1935, and Thursday, February 18, 1937, p. 22.
27. Emporis. Woolworth Building, available at: http://www.emporis.com/building/woolworthbuilding-newyorkcity-ny-usa.
28. Emporis. Willis Tower, available at: http://www.emporis.com/building/willis-tower-chicago-il-usa.
29. http://www.asianoffbeat.com/post/Six-of-the-World%27s-10-Fastest-Elevators-Are-in-Asia-news-980.
30. Fosburg, P.L. Elevators. In: *Gale Encyclopedia of US History*, extract available at: http://www.answers.com/topic/elevator.
31. Harrison, B. *Seeking a Role: The United Kingdom 1951–1970*, p. 158.
32. University of Coimbra. E4 Project, Energy Efficient Elevators and Escalators, available at: www.e4project.eu/documenti/wp6/E4=WP6-Brocure.pdf; the figures are actually for 2010.
33. National Elevator Industry Inc, Elevator and escalator fun fact, available at: www.neii.org/presskit/printmaster.cfm?plink=NEII Elevator.
34. Otis. About elevators, available at: http://www.otisworldwide.com/pdf/AboutElevators.pdf.

20

Gadgets: Small Household Appliances

Although a food processor is not an absolutely essential piece of equipment, because you can certainly chop, grate, slice, knead and mix everything by hand, it does do all these things very quickly and efficiently and saves you time and energy.

Delia Smith, cookery writer

Most early electrical appliances depended on the heating effect of electricity. Cookers, kettles, fires and toasters were all a means of supplying heat in the appropriate manner. It was the controllability of electricity and its ability to supply that heat in a precise and efficient manner that gave these items their popularity. However, electricity could do a great deal more than this—it could produce motion, as shown by the many railways that were electrified. The problem was in trying to make the motors small enough for domestic use.

Curiously, one of the first appliances to appear was the electric fan in the early 1880s, but it was not really practicable for the average household, even if it had an electricity supply.[1] Even by the end of the century, though the blades were quite efficient, the motors were still large and open and, hence, dangerous and inefficient.

In 1907, the split ball joint was added to the design which allowed the fan to oscillate from side to side and so blow air over a wider area.[2] This progressed faster in America because electrification came earlier and many parts of the country had higher temperatures, and hence a need for cooling, than in England. It only needed improvements in the motors for the fan to be a satisfactory device.

In 1910, in Racine, Wisconsin, businessman Fred Osius hired ex-steamboat cashier, L.H. Hamilton, to do his advertising with yet another former farm boy, Chester Beach, as the technical man. Beach was to come up with a small "universal" motor than ran on both AC and DC, which was a considerable advantage when both types of supplies were common in homes.[3]

It is often claimed that Beach invented the universal motor, but it was well known years before. The commutating AC motor, which would of course run on DC, was being used for trains as early as 1904.[4] The motor is not as efficient on AC as on DC due to the inductance

J.B. Williams, *The Electric Century*, Springer Praxis Books, DOI 10.1007/978-3-319-51155-9_20

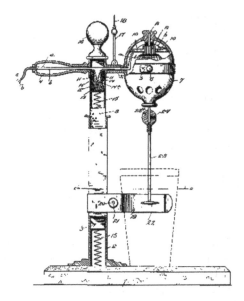

Fig. 20.1 Fred Osius' liquid beverage mixer using Beach's motor. Source: US patent 582,863.

causing the current to lag the voltage and so the power not being as high as it would be on DC.[5] For heavy traction motors this is important and so frequencies at low as $16\frac{2}{3}$ Hz were used to minimize this effect. However, for small motors it doesn't really matter. Beach's contribution was to make the motor small enough for small appliance use. Fred Osius used the motor as part of his patented drinks mixer (Fig. 20.1).[6]

There must have been something about Racine, Wisconsin, because the next step was also taken there. Stephen Poplawski came up with the design of a drinks mixer with blades mounted in the base of the container—what is now called a blender.[7] In 1922, he patented the device and assigned it to the Arnold Electric company for whom he worked.[8] The machine was intended for soda fountains and particularly for mixing Horlick's Malted Milk because that company was also based in the town.

Meanwhile, Fred Osius, not wanting to use his own name, and having called his company Hamilton Beach after the two employees, lost them when they left to form their own Wisconsin Electric Company. Despite this, Osius continued to build up the company before selling it to the Scoville Corporation. However, he still remained interested in blenders and in 1937 filed for a patent for an improved device.[9]

Seeking a way to promote it and finance its development he approached the well-known band leader, Fred Waring. It seems a strange choice, but Waring agreed to back the project. Unfortunately for Osius this soon backfired, as Waring realized that Osius was taking too long and that the device still had technical issues. Osius was dropped and the device redesigned. The product that was produced was known as the Waring Blendor—the "o" was to give it class—and it soon was the must-have device for mixing frozen daiquiris in bars. It was to go on to have another use in laboratories, most famously that of Dr. Jonas Salk who produced the polio vaccine. By 1954, Waring had sold a million of his Blendors.[10]

There was one more step in this tangled web. In 1926, Arnold Electric Company was sold to Hamilton Beach Manufacturing Co. and Poplawski ended up working for them. He could see that there was a potential market for a home blender and worked privately on a suitable device. He went on to form the Stephens Electric Co., and in 1940 he patented a household mixer for the home.[11] In 1946, this became the "Osterizer" when Poplawski sold his business to the John Oster Manufacturing Co.

Though many inventors started to look at small appliances and some were produced for commercial applications, it was only really after World War I as household electrification took hold that real household appliances began to appear. The Hobart Manufacturing Company of Troy, Ohio, had been making mixers, not for drinks but for kneading dough in bakeries since 1915, but in 1919 produced their H5 which was for making cakes and the like in the home.[12] It used a "planetary action" whereby the beaters rotated in one direction while the mix was moved around the bowl in the opposite direction.[13] This was the Kitchen Aid mixer.

In 1930, the Kitchen Aid mixer attracted serious competition from the unlikely named Chicago Flexible Shaft Company which made tools for grooming farm animals. They had employed a young Swede, Ivar Jepson, as their designer[14] and he produced the Mixmaster which was sold under the Sunbeam name. It was simpler and cheaper than the Kitchen Aid and by 1936 300,000 Mixmasters a year were sold in the worst of the Depression.

Britain was much later into mixers, as with most household electrical equipment, and before the war there was only a small market, mostly for American imports. However, in 1948 Kenneth Wood set up a manufacturing company for domestic appliances. Two years later, at the Ideal Home Exhibition he launched the Kenwood Chef. Not only did it have a "planetary action", but it came with a beater, a whisk and a dough hook for making bread. That wasn't all; a range of extras were available for many other kitchen tasks where it could replace the blender or mincer or juicer, and also slice, chop and peel.[15] This made sense when the motor was the expensive part of the machine, despite the inconvenience of having to fit and remove the various tools.

This early model was rather industrial looking but was acceptable in the 1950s. For the 1960s, Ken Wood employed another Ken, Kenneth Grange, to restyle the Chef; he replaced the rounded curves with a squared-off design while retaining the mechanics.[16] From that point on the Chef has changed little and is still a feature of many kitchens.

These starting points led to numerous models from other manufacturers, in particular simpler hand-held mixers suitable for the household where a relatively small amount of baking took place. By the middle of the 1980s, perhaps as many as three-quarters of homes in Britain had a mixer, even though the amount of home baking was in serious decline.[17] It had become a normal piece of essential kitchen equipment.

In France, a catering company salesman, Pierre Verdon, found that many of his clients spent a lot of time chopping, dicing, grinding and blending ingredients in their kitchens. Though some of these could be undertaken either by blenders or the attachments to the mixer, nothing efficiently handled dry materials. To save the vast amounts of time that these operations took, he invented the Robot Coupe which, unlike the traditional blender, had a shallow wide bowl.[18]

Eventually he produced a home version, le Magi-Mix, which appeared in the trade shows in 1971 and on the market in 1972. Carl Sontheimer, an American engineer raised in France, spotted the Magi-Mix at one of the French trade shows. He realized its potential and, together with his wife Shirley, founded the Cuisinart company to bring European cookware to the US. Working with Verdon, he redesigned the machine by adding safety mechanisms and improving the chopping disks and blades. The new machine was launched in 1973, and though sales were slow to start with, it eventually took off after celebrity endorsement.

While there is some overlap between blenders, mixers and food processors, each has its strengths. It depends on what the cook likes making to decide which is the best appliance for them. Numerous companies have offered products into these markets and they became the staples of kitchens everywhere for the rest of the century. Many cooks would never consider working without their favorite helper.

One of the everlasting tasks in the kitchen is washing up. It is hardly surprising that there have been numerous attempts to make machines to undertake this chore, stretching back into the nineteenth century. However, it would take the coming of electricity, and particularly the electric motor, before a satisfactory device was produced. At first sight it would seem that dishwashers don't belong amongst small appliances, because nowadays they are large, but they didn't start that way. In 1924, Englishman Howard Livens came up with a design that used water pressure to drive arms which sprayed hot water over baskets containing the dirty dishes.[19] It wasn't actually an electrical machine, but it did contain most of the elements of a modern dishwasher, including a front door for loading. In principle, it was suitable for the domestic environment, but it didn't really catch on probably because it didn't appear as a neat and simple machine.

Soon makers of other appliances began to look at this market. General Electric in America produced a machine in around 1928, while Miele in Germany launched theirs in 1929. It might have been thought that there would have been considerable interest, but it wasn't so. In Britain, there were problems with the supply of electricity but not of dishwashers (human and usually female), while even America with its greater penetration of electric supplies, and generally a shortage of domestic labor, still didn't equal a large market.

The manufacturers steadily improved their machines, and after World War II when affluence brought a greater interest in household equipment, the sales began to rise. The rise was faster in America but what was surprising was that the machines were never as popular as might have been expected with only a million sold by 1951.[20] By the end of the century only some 50% of homes had them (Fig. 20.2). In Britain, they only really began to take off around 1980 and then had only reached around a quarter of households by the end of the century.

At first sight, this lack of popularity is surprising. The machines are relatively simple compared with a washing machine. There is no rotating drum and only a pump is required to spray the water out of the nozzles and over the dirty dishes. A second pump removes the dirty water when the cycle is complete. These, together with a valve to let the water in, a heater for achieving the right temperature and a timer to control the sequence, are all that is required. Thus the machine can be cheaper than a washing machine which is in virtually every household.

Curiously, the penetration in Britain is lower than other major European countries and even in the US little more than half of the homes have one.[21] The industry itself really

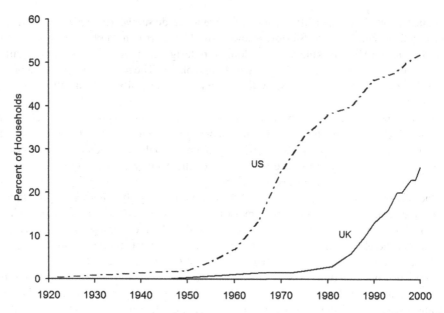

Fig. 20.2 The low take-up of dishwashers in the US and the UK, 1920–2000. Source: Author.[22]

doesn't know why. The reasons put forward are: it is not considered essential like a washing machine or refrigerator; it consumes a lot of water, whereas the modern ones don't; and that it takes up a valuable cupboard space in the kitchen.

None of these are very convincing. It is true that equipping the man of the house with a pair of rubber gloves and some washing-up liquid is much cheaper, but it still doesn't offer a real explanation. People just don't regard a dishwasher as essential. It is interesting to note that while the penetration varies between advanced countries, with Britain at the bottom and the US at the top, nowhere does it reach the levels of many other appliances.

It wasn't only in the kitchen that the coming of the small electric motor had an impact; there were other items in the home. An obvious one was an applicance to dry hair after it had been washed. Left to itself, long hair can take a very long time to dry and to speed this up some combination of heat and air movement is required. Clearly, a stream of hot air would do the job.

It is often stated that, in the 1890s, the hot exhaust from vacuum cleaners was used to dry hair. This is, of course, nonsense as the vacuum cleaner wasn't invented until well into the twentieth century (see Chap. 12). A practical hairdryer requires a heating element and to be a reasonable size this needs the introduction of Nichrome wire, and also a small electric motor to drive a fan. Thus it was after World War I when these things all came together and the first hairdryers that we would recognize started to appear.

The fundamental design of the hairdryer has changed little since. A motor driven fan — usually tangential — sucks in air from the side or rear and blows it out through a nozzle where it passes over a heating element. The main difference is that these early models were made of metal and were very heavy. In addition, the heating elements were only around 100 W and so they were rather slow to dry the hair.[23]

In the 1950s, the introduction of plastic cases led to lighter, cheaper devices, and steadily the power of the heater was increased to 300 or 400 W. Around that time there was a fashion for curly hair and the hair dryers were supplied with a plastic hood (often pink) which was placed over the hair and the user could sit comfortably while the hair dried.

In succeeding years this fashion disappeared and the power was increased to 500 W, then 1000 and by the end of the century it was well beyond that, giving much quicker drying. Obviously, this can only be achieved with the introduction of a number of safety devices to prevent it from overheating. The hairdryer, while appearing to be the same device, has become more sophisticated to pack more power into a smaller and lighter size. Almost all homes have some sort of hairdryer.[24]

With such a large potential market, it was inevitable that the electrical appliance industry would have a go at an electric shaver. However it required an even smaller motor for something that could conveniently be held in the hand. Though Jacob Schick applied for a patent for his device in 1928, it was not until the late 1930s that real commercial devices appeared.[25] One of the first was from the Remington Rand Corporation in 1937, but they were soon joined by other manufacturers in Europe such as Siemens and Rolls Razor.

In 1939, engineer Alexandre Horowitz came up with the design of the Philishave which used a rotating cutter instead of the reciprocating ones used by other manufacturers.[26] Though the Philips company had to stop production when the Netherlands was occupied during the war, they continued with the product once it was over. The original devices had a single cutting head, though later they went to two and even three heads.

Despite the successful designs and the convenience of not having to lather the face and the other complications of wet shaving, electric razors have never taken over completely. Some of this is conservatism of the users, but many think they get a better result wet shaving and inevitably it is a matter of personal preference. Despite the fact that they can be used by both men and women, not every household has one. In the mid-1980s this was still at around 44% of households.[27]

It was inevitable that electrical gadgets should try to conquer other areas of personal care, and the toothbrush was an obvious target. The reasoning was that many people are not sufficiently energetic with their manual brush and hence a motor-powered one would be an advantage. The first electric toothbrush was invented by a Swiss dentist, Dr. Philippe-Guy Woog, who called his device the Broxodent and it was initially aimed at patients with limited motor skills or those wearing braces on their teeth.[28]

The disadvantage of Woog's device was it ran from the mains, and hence had safety concerns. This problem was addressed in 1961 by General Electric in America who produced a small device containing a rechargeable battery.[29] It could be placed on a stand to recharge but was disconnected from the mains when in use, hence producing a much safer device. This was the solution and sales began to increase.

There was one drawback with the GE device: because of the size of the rechargeable batteries it was rather large. It took until the 1970s for this situation to improve when other companies such as Oral-B, Braun and Philips introduced improved devices based on even smaller motors.[30] Though electric toothbrushes became quite popular, only a small minority of households, something like 3% in 1985, had one or more.[31] A manual tooth-brush is much cheaper, and all it requires is a bit more effort.

The items mentioned above are by no means all the electrical household gadgets. We could mention electric carving knives, Teasmades, coffee grinders and coffee makers, hostess trolleys and a host of other devices. They all depend on heating elements and/or small electric motors. Though none of them have become essentials in the home they nevertheless still sell in numbers attractive to the manufacturers. Without these devices the modern home would not be the same.

NOTES

1. Cunningham, S. A brief history of fans, available at: http://www.fancollectors.org/fanhistory. htm.
2. Patterson, J.M. Oscillating fans—History and advancements, available at: http://ezinearticles. com/?Oscillating-Fans---History-and-Advancements&id=2627119.
3. Pollard, J., The eccentric engineer. *IET, Engineering & Technology*, 7:11, 104, 2012.
4. Creedy, F. Alternating-current commutator motors. JIEE, 33:169, 1163–1175, 1904.
5. The instantaneous power is the voltage times the current at that moment, so if they are not in phase (here the current is delayed) then the power out is reduced. However, the losses are defined by the actual current and voltage and so are the same as on DC. The result is less power out for the same losses and so a lower efficiency.
6. US patent 1,005,653.
7. Wisconsin Historical Society. Dictionary of Wisconsin History, Poplawski, Stephen J. 1885–1956, available at: http://www.wisconsinhistory.org/dictionary/index.asp?action=view&term_ id=12456&term_type_id=1&term_type_text=People&letter=P.
8. US patent 537,439.
9. US patent 2109.501.
10. Wikipedia. Fred Waring, available at: http://en.wikipedia.org/wiki/Fred_Waring.
11. US patent 2304.476.
12. Idea Finder. Mixer, available at: http://ideafinder.com/history/inventions/mixers.htm.
13. Funding Universe. Kitchen aid history, available at: http://www.funguniverse.com/company-histories/kitchenaid-history/.
14. MIT. Inventor of the week, Ivar Jepson, available at: http://web.mit.edu/invent/iow/jepson.html.
15. Watson-Smyth, K. The secret history of: The Kenwood Chef A700, *The Independent*, January 28, 2011.
16. Braggs, S. Kenwood Chef, available at: http://www.retrowow.co.uk/retro_collectibles/60s/kenwood_chef.php.
17. A figure of 70% comes from averaging a number of surveys conducted by schoolchildren as part of the BBC's Doomsday Reloaded project—entry point at: http://www.bbc.co.uk/history/ domesday. However adding the figure for different types of mixer in *Euromonitor*, The small electrical appliance report, p. 8 gives 76%.
18. Butler, S. Natural history of the kitchen: Food processor, available at: http://www.eatmedaily. com/2010/06/natural-history-of-the-kitchen-food-processor/.
19. British patent 219,103, 1924.
20. Hardyment, C. *From Mangle to Microwave*, p. 154.

21. *KBB Review.* Market analysis: Dishwashers, available at: http://www.kbbreview.com/Home/ market_analysis_dishwashers.htm; Veetsan Machinery, The development of the dishwasher, available at: http://veetsan.com/en/newsDetail.asp?id=29.
22. Data from: UK: Science Museum, Making of the modern world, available at: http://www.makingthemodernworld.org.uk/stories/the_rise_of_consumerism/02.ST.03/?scene=3; and ONS, General Household Survey; US, R. McGrath, The pace of technology is speeding up. *Harvard Business Review*, November 25, 2013, available at: https://hbr.org/2013/11/the-pace-of-technology-adoption-is-speeding-up/.
23. *Lifestyle Lounge.* History of hair dryer, available at: http://lifestyle.iloveindia.com/lounge/history-of-hair-dryer-6804.html.
24. It was already 76% in 1985. *Euromonitor*, The small electrical appliance report, p. 8.
25. US patent 1,757,978.
26. Wikipedia. Philishave, available at: http://en.wikipedia.org/wiki/Philishave; Byers, *The Willing Servants*, p. 48.
27. *Euromonitor.* The small electrical appliance report, p. 8. Curiously, the BBC Doomsday Reloaded project—see above—gives a figure around 70% which suggests that they were perhaps working in more affluent areas.
28. Wikipedia. Electric toothbrush, available at: http://en.wikipedia.org/wiki/Electric_toothbrush.
29. General Electric Appliances. History of appliance innovation, available at: http://pressroom.geappliances.com/historical/history-of-appliance-innovation.
30. Oral-B. History, available at: http://www.oralb.com.mx/lt/aboutus/history.asp.
31. *Euromonitor.* The small electrical appliance report, p. 8.

21

Freedom of the House: Central Heating and Air Conditioning

Central heating also meant the rise of the bedroom as a living—not just sleeping space—for children, rather than everyone congregating in the living room or kitchen.

Eleanor John, Head of Collections and Exhibits
at the Geffrye Museum, London

On Friday December 5, 1952 passengers in a train traveling from Aldershot to London in the UK wondered why it was getting slower and slower and even stopping periodically. Looking outside they realized that there was a thick fog. When they eventually reached Waterloo the guard had to come along the train hammering on the doors to persuade them that they were in the station as they couldn't see the platform and it was an act of faith to step out.[1]

Into the evening the fog became even denser and was described as "an old fashioned pea-souper, thick, drab, yellow, disgusting".[2] Over the next few days road traffic virtually stopped as progress could only be made if someone walked in front to guide the way. Even then there were accidents and ambulance men and firemen had to walk ahead of their vehicles to reach the scene.[3] Theaters had to close as the audience couldn't see the stage, and sporting events were cancelled due to the impossibility of playing or viewing the action.

The weather had been very cold and the conditions had been such that warm air settled over the top forming what is known as an "inversion". This meant that people piled more coal on their fires to try to keep warm but all the smoke was trapped under the inversion layer and couldn't escape.[4] The result was the worst smog that had ever been experienced in London, with 4000 people dying soon after it and it is now thought that a further 8000 did so subsequently.[5]

Though London was probably the worst city for these fogs due to its size, it was a common problem. Most domestic heating in developed countries was by burning coal, either in open fires, stoves or, in America, quite commonly in furnaces. This was only the beginning of a growing realization around the world that the mass burning of coal in cities couldn't go on. In Britain, this incident brought in the 1956 Clean Air Act forbidding the burning of dirty fuel in the cities.[6]

© Springer International Publishing AG 2018
J.B. Williams, *The Electric Century*, Springer Praxis Books, DOI 10.1007/978-3-319-51155-9_21

The Act concentrated on dust and grit but the measures also had the effect of reducing the sulfur dioxide which was probably the real killer. Though progress was slow at first, this marked the beginning of the end of coal as the mainstay of domestic heating. For centuries, the home fires had had an almost mythical place in the consciousness of the population. At the end of World War II some 95% of homes used a coal fire to heat their living room.[7] The hearth was regarded as the center of the home; now it was under attack.

The question was what to use to replace it. Some people opted to burn more expensive smokeless fuels, but this required alterations to the grates, though there was some financial support to make the change. However, for many others the possibility of getting rid of the labor of carrying coal in and ash out, as well as the dust and dirt associated with it, seemed very attractive.

America was fortunate in that, during the 1940s, gas pipelines had been built from the oilfields in the South West to the Mid West and East.[8] This meant that in the period of rising prosperity postwar another fuel was available to replace coal, which it did quite rapidly. Also with the supplies of domestic oil meant that was another option. Figure 21.1 gives the decline in coal use in Britain, but Fig. 21.2 shows that it happened earlier in the US due to the availability of the alternative fuels of gas and oil.

Most of the gas was piped to homes, but tanks or bottles were used where a mains supply was not available. This was the main alternative fuel to coal, through until 1960 when price hikes made it less economic. Oil, or kerosene, was a popular option, particularly in the countryside or in small towns which the gas mains didn't reach. The surprise is that as the use of gas topped out and oil declined, it was mostly electricity that took over. It all depended on local conditions. In some cases gas was relatively expensive and electricity

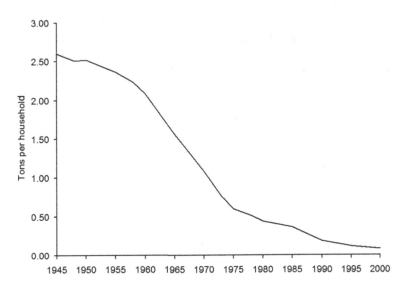

Fig. 21.1 The decline in the average annual household use of coal in Britain, 1945–2000. Some homes still used solid fuel but their numbers became fewer and fewer. Source: Author.[9]

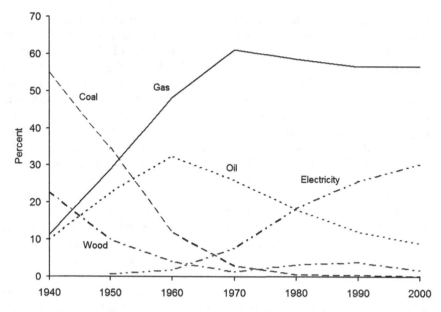

Fig. 21.2 The changing US domestic fuel mixture, 1940–2000. Source: Author.[10]

cheap, and in other areas the reverse was true, and this drove the take-up of the different fuels. Many of the electric heaters were built in either as panels fixed to the wall or as baseboard heaters along the bottom of the wall. This can set up a convected current of warm air up the wall and hence heat the whole room.

Britain didn't have the options of gas or oil. At the time, all oil had to be imported and gas had to be made from coal which was relatively expensive. There was no convenient domestic source of natural gas. However, there was one solution which had been popular since the end of the war—the electric fire.[11] Despite the government's efforts to curtail the sales of these during the fuel shortages of the late 1940s they were easy to install and use. It only required plugging in and most houses now had the requisite power sockets. A flick of the switch was so much easier than lugging buckets of coal, but the electric fire still lacked something as the focus of the living room.

However, that situation was changing as the television set steadily took over that role, and the fire as the center of the home became less important. In fact, in the layout of many small rooms there was a considerable conflict in the positioning of the TV and the existing fireplace. In millions of homes the chairs were now positioned for viewing rather than to be near the heat, though some compromise needed to be made in winter.

Many electric fires were used to heat bedrooms when people were getting up. Unfortunately, this tended to coincide with the morning peak of electricity demand, and was a considerable worry to the electricity supply industry.[12] From the mid 1950s, because of the shortage of generation capacity, there were moves to try to persuade consumers to use "off-peak" electricity at night or at other slack times such as the middle of the afternoon. The bait was cheaper prices, but despite some installations of underfloor heating, this didn't attract a great deal of interest.

In 1961, the Electricity Boards, almost uniquely in Britain, started marketing domestic storage heaters. These were wired on a separate circuit which was switched on at the "off-peak" times by a time-clock. This supply was metered separately by a second meter and charged at around half the normal price.[13] The heaters were made with large heavy bricks inside (some used tanks of water) which heated up when the electricity was switched on and then gave it out steadily during the time when the power was not available.

While not true central heating the heaters could give out a background heat which kept the room warm all the time. The system also had the advantage that heaters could be added one or two at a time, when they could be afforded, until the whole house was covered. The disadvantages were that the heaters were big and bulky and in cold weather they often ran out of the charge of heat during the day even with the afternoon boost. They gave out most heat at night when it was not particularly required, but they did generally keep the house nice and warm.

Throughout the 1960s night storage heaters became quite popular due to their flexibility and low installation cost. The numbers steadily increased until 1975 when about ten percent of homes had them as their main form of heating.[14] After that the numbers fell slightly, as electricity prices reacted to the fuel shocks of the early 1970s. From there, for the rest of the century, the numbers slowly increased, but so did the number of households so the percentage remained roughly constant.

The gas industry could also offer fires. They were mostly piped into a fixed position but there were some with a connector so that they could be disconnected and moved to another room. Though not quite as convenient as electric fires, their burning flames made them more suitable for the main rooms of the house.

Just after the war, with electricity somewhat underpriced, gas fires were not markedly cheaper to run than electric fires, but this gradually changed as electricity prices became more realistic. The problem with gas was that it was made from coal and only extracted part of the available heat and so was relatively expensive to produce. The industry was also inefficient with large numbers of small gas works, some of them very old.

Despite attempts to introduce other sources with products derived from natural gas, it could not really make an impact on the home heating market. This was about to change when in 1965 natural gas deposits were discovered in the West Sole Field in the North Sea.[15] The following year a drastic decision was made to convert the whole country to natural gas. The program started in 1968 and took around 10 years to complete.

Town or coal gas is mainly hydrogen and carbon monoxide (which is why it is poisonous) but natural gas is predominately methane; the burning characteristics are different so the burners needed to be changed on appliances. The task was to convert 40 million appliances from 14 million consumers and was only justified by the enormous savings that the industry would make when it no longer had to manufacture gas from coal or try to convert other gases to be compatible with it. The pipeline already built for the imported natural gas provided a starting point for the distribution of the North Sea gas and eased the conversion.

The change transformed the prospects of the gas industry which had been losing out to electricity in almost every area. Between 1963 and 1973 the effective price of gas dropped by around a third while that of electricity didn't change much. As the natural gas started to flow, consumption took off, trebling between 1969 and 1977. Gas had arrived as a serious heating fuel.

Some forms of central heating and been around since the late nineteenth century. It usually consisted of a large coal-fired boiler or furnace in a basement. The system needed someone to keep the boiler or furnace fed with fuel and adjust the dampers from time to time to keep the temperature for the occupants within a reasonable range. There were a number of ways that the heat could then be distributed around the building: steam, water or air.

The water or steam systems depended on the fact that hot water or steam is less dense than cold and rises to the top. Hence, the radiators needed to be above the boiler so that the heat rose to them. As the water in them cooled it would sink back in the return pipes to be heated again. The flow depended on this thermal movement and was known as a "gravity" system. To get an adequate flow of heat the pipe diameters needed to be quite large.

In America, where basements in houses were common, it was easy to site the furnace there. For some reason, probably due to the activities of some entrepreneurial suppliers, hot-air systems became popular.[16] These again depended on the hot air rising. In the basement was the furnace, often known as the "Octupus" because a series of massive ducts sprouted out of the heating block to lead the heat up into the various rooms.[17] It also depended on the hot air rising.

Clearly these systems were only suitable for particular types of buildings. They were not going to work well in the average house which had no basement and most of the heat was needed on the ground floor where the boiler or furnace also had to be sited. The solution for the hot-air systems was to introduce electric fans to blow the heat around, and for the water systems a pump for the same purpose. Steam systems only really were used in large buildings and their dangers soon led to their demise.

Once satisfactory supplies of natural gas were available it was an obvious step to replace the boiler or furnace with a gas-fired one. For those without the gas supply, oil was an alternative. It was at this point that the gas industry had to call on their great rival, electricity, for help. To provide a satisfactory heating system there needed to be some automatic method of controlling temperatures in the building. It was not easy to adjust the output of the gas boiler with dampers in the same way as coal-fired furnaces. A simple gas boiler or furnace is either on or off.

The solution was to use a thermostat suitably positioned in the building. When the temperature fell the contact on the thermostat closed and an electrical signal was sent to open a gas valve supplying the boiler or furnace. This would be lit from a small continuous flame, or pilot light, and the boiler or furnace would roar into action, heating the house until the thermostat registered that the temperature was high enough and opened its contact causing the gas valve to close and the heat to shut off.

In Britain, the hot-air systems had a period of popularity in the 1960s. The numbers installed each year steadily increased until they reached a peak in the later part of the decade, but then started to decline. All together, more than half a million homes had them by 1970.[18] After a time, they were often found to be noisy and could produce drafts. Unless filters were used and cleaned regularly dust and dirt would blow around the house. They largely fell out of use, which is curious as they were by far the most common system in the US.

The original problem with hot water systems had been installation cost. Much of this was due to the iron pipes used. The ends of each section had to be laboriously threaded by hand using a die, and couplers screwed on. It was extremely time consuming and required skilled installers, but there was a solution. Once a pump was used, the pipes could be of smaller diameter and this opened up the possibility of using standard plumbing copper pipe which can be easily soldered.

The bulky radiators now didn't need to be so large and could be replaced by simpler and neater designs made from pressed sheet steel and welded. These advances, together with smaller and lighter boilers, produced a system that was much cheaper to install and flexible enough for most houses. The pump was powered at the same time as the gas was turned on and controlled by the thermostat.

The result was that these hot water systems rapidly started to take off in Britain in the 1960s. They were being installed at the rate of some 300,000 a year by the end of the decade and by which time some one and a half million were in use. From then on the number climbed ever upward as the idea that central heating was achievable for the average home became widespread. There was another improvement, again calling on electricity. Traditionally, many households had damped down their coal fires at night or when they were out so they could be brought to life late without too much trouble. The central heating system didn't provide for that, but adding a timer meant it could be arranged to switch on and off at preset times. At first, these were simple electromechanical timers, but they were later replaced by electronic ones that could do more sophisticated things such as have different timings at the weekends or on different days of the week. Now a thoroughly practical system had been produced (Fig. 21.3).

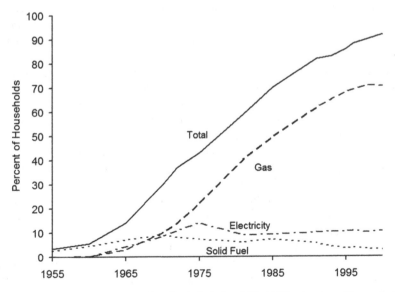

Fig. 21.3 The rise and rise of central heating in Britain, 1955–1995—dominated by gas boilers and water-filled radiators. Source: Author.[19]

There were still improvements that could be made. The boilers or furnaces were still rather large and, being floor mounted, took up what could otherwise be a useful cupboard or appliance space in cramped kitchens and utility rooms. By careful design the units were made still smaller and lighter which meant that they could be hung on the wall up out of the way. This increased their acceptability still further.

As the century marched on, and fuel started to become more expensive, attention turned to the efficiency of the boilers or furnaces, which was not very high. Some couldn't even convert 60% of the energy in the fuel to useful heat output. The first thing to do was to get rid of the constantly-running pilot light. As the boiler or furnace was now controlled by electricity it was not difficult to introduce an electronic lighter which generated a high voltage spark to ignite the gas at the moment when the thermostat demanded heat.

A further improvement was the addition of an electric fan which could blow the air and gas mixture into the burner. The advantage of this was that the output power could be adjusted by altering the speed of the fan while still maintaining the correct gas-to-air ratio. It required an electronic control system to keep the temperature of the hot water or air in the heating system at a constant level while it was switched on. This was considerably more effective than simply switching the heat on and off.

To gain even more efficiency, heat exchangers were mounted in the flue which extracted a greater proportion of the heat from the burning gas. This was so effective that the temperature of the flue gas was low enough for the water vapor produced in the combustion to appear as liquid water. It was thus necessary to have a drainpipe to lead this away. This type was known as a condensing boiler or furnace because of the water condensed out of the flue gas.

Achieving this performance—and these could reach efficiencies above 90%—needed very careful sequences of operation. For example, it was essential to ensure that no residual gas remained in the burning chamber when it was switched off otherwise an explosion could occur. The solution was to have yet more electronics in the form of a microprocessor which could implement all the complex operation sequences and check all the interlocks to prevent incorrect operation.

Thus a heating system that uses gas, or even oil, as its primary fuel utterly depends on electricity for its operation, as anyone who has tried to stay warm in a power cut will have discovered. Thus electricity, which—rightly—largely lost the battle for central heating (though less so in America), crept back in for its command and control functions. There is no avoiding its ubiquity. Without it, central heating would never have become a practicable system.

Thus, by the end of the century central heating had become the norm, even in countries like Britain. As a result, the whole house was kept at a usable temperature even in the depths of winter. This meant that it was no longer necessary for everyone to pack into the one heated room trying to avoid the draughts. Schoolchildren could go somewhere quiet to do their homework, and annoying teenagers could hang out in their bedrooms to play the music that was alien to their parents.

It is sometimes argued that this helped to break up the family, with parents and children not all being together around the hearth.[20] This forgets that in summer this wasn't necessary. In any case, the rise of the teenage generation happened in the 1950s and 1960s, well before central heating became general. On balance, central heating has brought positive benefits. Virtually no one would want to go back and try to do without it.

In warmer climates, in the south, the problem is not so much keeping warm as staying cool. The answer was there in refrigeration, but the question was: how to apply it to buildings? Willis Haviland Carrier was another American farm boy, but he trained in mechanical engineering and ended up specializing in heating systems.[21] In 1902, he was called to try to improve the conditions in the Sackett-Wilhelms Lithographing & Publishing Company of Brooklyn who were having difficulties with the temperature, and particularly the humidity, in their printing shop.

He soon realized the complexity of the problem, and set about producing a system that not only cooled the air but also adjusted the humidity by either adding water with sprays, or drying it to produce air not only at the right temperature but also with the humidity in a comfortable range. The customer was delighted with the equipment which stopped the ink running. Carrier's success was down to a thorough understanding of the relationship between humidity and temperature. In 1904 he patented his ideas.

The equipment to do this was quite large but it was suitable for factories, shops, stores and other large buildings, and his company, formed in 1915, had a steady business in these public buildings. Gradually, the systems were made smaller and lighter until after World War II simplified systems were suitable for use in houses. In the early 1950s, riding on the building boom, significant numbers of air conditioning units began to be installed. The heart of these was a cooler unit based on refrigerator technology, but it was utterly dependent on electric pumps and fans.

In American homes, the air central heating systems made it possible to integrate the air conditioning and pass cool air around the house as well as hot. Obviously, these are more common in the hotter regions in the south, but they are quite popular over the whole of North America (Fig. 21.4).[22] They are almost unknown in homes in Britain and in Europe, though they are present in significant numbers in public buildings.

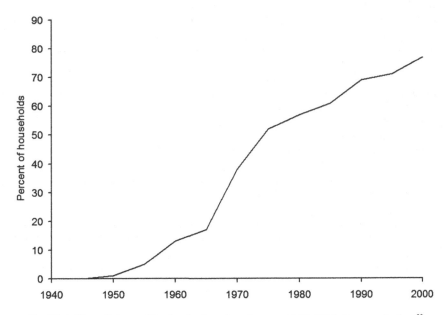

Fig. 21.4 Rise of air conditioning in American homes, 1940–2000. Source: Author.[23]

There is another form of air conditioner which is even simpler and intended for just one room. Typically the "hot" unit is outside a window on the wall, while the "cool" unit is inside, and the two are connected by pipes carrying the refrigeration fluid. This form is common in older buildings in America, particularly further north. They have spread round the world in hot climates, but in Britain they are rare except in offices and other commercial buildings.

Strictly, all these domestic units are not true air conditioners, as they only seek to control temperature and not humidity as well. Only the large building units do this. However, they still make an enormous difference to the living conditions. In America, the rise of air conditioners has caused a general population shift to the "sun belt".

It is possible, of course, to run the unit in reverse. There is no basic difference between the hot and cold units. By rearranging the pumps and valves the unit can be made to heat the building rather than cool it. With careful design it can do either depending on the weather conditions. In this form it is known as a heat pump. Sometimes, this is taken even further and the outside unit is sunk in the ground to provide the heat source or sink. These ground source heat pumps are very efficient, but they have never become very popular, probably because of the amount of space needed for the pipes in the ground.

Thus the environmental control of buildings, either public or the home, which we now take for granted is only possible because of electricity. It may not always provide the basic heat source, but it pumps, fans and controls it in such a manner that it would not be practical without it.

NOTES

1. Kynaston, D. *Family Britain 1951–57*, p. 255.
2. *The Guardian*. Our London correspondence, December 6, 1952.
3. BBC. On This Day, December 9, 1952: London fog clears after days of chaos, available at: http://news.bbc.co.uk/onthisday/hi/dates/stories/december/9/newsid_4506000/4506390.stm.
4. Met Office Education, The great smog of 1952, available at: http://www.metoffice.gov.uk/education/teens/case-studies/great-smog.
5. Bell, M.L, Davis, D.L. and Fletcher, A. A retrospective assessment of mortality from the London smog episode of 1952: The role of influenza and pollution. *Environmental Health Perspectives*, 112:1, January 2004.
6. Hansard, *HC Deb, January 25, 1955 vol 536 cc38–42;* Details of the Act available at: http://www.legislation.gov.uk/ukpga/Eliz2/4-5/52/enacted.
7. Political and Economic Planning. Report on the market for household appliances, p. 123.
8. Barreca, A., Clay, K. and Tarr, J. Coal, smoke, and death: Bituminous coal and American home heating, 1920–1959, available at: http://www.american.edu/cas/economics/upload/BarrecaClayTarrCoalStanford-paper.pdf.
9. Data from: 1945–1965: Mitchell, B.R. *British Historical Statistics, Fuel and Energy*, pp. 258–259; 1970–2000; Department of Energy and Climate Change, *Digest of United Kingdom Energy Statistics* 2.1.2 Inland consumption of solid fuels; Households extrapolated from censuses.
10. United States Census. Census of Housing, House heating fuel, available at: https://www.census.gov/hhes/www/housing/census/historic/fuels.html.

11. See Chap. 13.

12. See Chap 13 for more about peak loads.

13. Hannah, L. *Engineers, Managers and Politicians*, p. 215.

14. Calculated from data in: Shorrock, L.D. and Utley, J.I. Domestic Energy Fact File 2003, BRE Housing Centre; Additionally, a small percentage used non-storage electricity as their main form of heating.

15. National Gas Museum. History of gas: A chronology of the UK gas industry, available at: http://www.nationalgasmuseum.org.uk/index.asp?page=history-13.

16. Energy Information Administration, 2001 housing heating tables, space heating tables, available at: http://www.eia.gov/consumption/residential/data/2001/hc/pdf/alltableshc3.pdf.

17. Roginski, K. Heating with an old octopus furnace. The Old House Guy, available at: http://www.oldhouseguy.com/heating-old-octopus-furnace/.

18. Calculated from data in: Shorrock, L.D. and Utley, J.I., Domestic Energy Fact File 2003, BRE Housing Centre.

19. Data from: Shorrock L.D and Utley, J.I., Domestic Energy Fact File 2003, BRE Housing Centre; Arapostathis, Set al. Governing transitions: Cases and insights from two periods in the history of the UK gas industry. *Energy Policy*, 52, 25–44, 2013; ONS, General Household Survey.

20. Geoghegan, T. What central heating has done for us. *BBC News Magazine*, Thursday, 1 October 2009, available at: http://news.bbc.co.uk/1/hi/magazine/8283796.stm.

21. Who Made America? Willis Carrier, available at: http://www.pbs.org/wgbh/theymadeamerica/whomade/carrier_hi.html.

22. US Energy Information Association. Air conditioning in nearly 100 million U.S. homes, available at: http://www.eia.gov/consumption/residential/reports/2009/air-conditioning.cfm.

23. The pace of technology is speeding up. Harvard Business Review, November 25, 2013, available at: https://hbr.org/2013/11/the-pace-of-technology-adoption-is-speeding-up/.

22

Power Tools and the DIY Revolution

We become what we behold. We shape our tools, and thereafter our tools shape us.

Marshall McLuhan, Canadian professor and philosopher

The great technology of the Victorian era was the use of riveted steel plates to make everything from ships to bridges. While there has been much attention on the rivets and the need to heat them and hammer them over before they cooled, the other side of the coin has been ignored. All those rivets had to go somewhere and that meant drilling thousands and thousands of holes into big slabs of steel. This was achieved by large fixed drilling machines which meant that the work pieces had to be brought to the drill rather than the other way around.

With the advent of electric motors it was inevitable that they would be applied to drilling machines, and in 1889 the Australian Arthur James Arnot invented and patented the first electric drill.[1] However, it was not portable and, apart from controllability, had few advantages as the work still had to be brought to it.

While this was satisfactory most of the time there were occasions when it would have been convenient to move the drill and make the holes in situ. This became a practical proposition with the flexibility of the electrical power connection. In 1895, the first hand-held portable electric drill was developed by the German brothers, Wilhelm and Carl Fein, in Stuttgart. It was a fearsome brute and had to be held by handles on each side and pressed against the chest to apply sufficient force to make the hole. The Fein brothers, however, went on developing their drill and by 1900 were using some aluminum parts to try to get the weight down.[2]

In 1910, two young American men, S. Duncan Black and Alonzo G. Decker, tiring of working for an unpredictable employer, set up a small engineering company of their own.[3] At first, they took in whatever work they could find but their ambition had always been to make their own products. By 1916, they had a patentable idea and that was to make a portable electric drill which could be used in one hand (Fig. 22.1).[4] It was loosely modeled on a Colt pistol; in place of the trigger was a switch which powered up the drill while it was pressed. It was launched in 1917 and became a roaring success with the company's annual turnover reaching $1 million by 1919.

© Springer International Publishing AG 2018
J.B. Williams, *The Electric Century*, Springer Praxis Books, DOI 10.1007/978-3-319-51155-9_22

Fig. 22.1 (*Left*) S. Duncan Black; (*Center*) Alonzo G. Decker; and (*Right*) their pistol trigger electric drill. Source: http://www.blackanddecker.ae/en/about/history/ and http://www.history.com/shows/modern-marvels/videos/the-worlds-first-power-tools.

This was the vanguard of the postwar revolution in portable power tools. Black and Decker continued to innovate, introducing a portable electric screwdriver in 1923, and also a lower-cost electric drill in the same year.[5] Other companies were also aware of the opportunities, with American inventor Raymond De Walt producing a radial arm saw in 1923.[6] The next year the Michel Electric Handsaw Company produced a circular saw under the trade name Skilsaw, which was originally intended for cutting sugar cane. Art Emmons of Porter-Cable, one of the most prolific inventors of power tools, developed the portable sander in 1926. That was soon followed by the jigsaw, invented by Albert Coffman working for what is now Bosch.[7]

Soon many companies around the world were making a whole range of power tools. In Britain in 1930. S. Wolf and Co. Ltd. were making drills, grinders, screwdrivers, and saws in a whole range of different types and sizes.[8] There were accessories for the drills for sanding, wire brushing, buff polishing and grinding. What all these products from the many companies had in common was that they were aimed at the commercial or professional market. Nobody considered that the average householder would want such things; they were the preserve of the skilled craftsman.

World War II was a good time for the power tool makers with a heavy demand from all the war production industries. Despite this, many of the companies were also involved in other war production. Black and Decker were doing brisk business with defense contractors and received the prestigious Army-Navy "E" award for their efforts. In Britain. S. Wolf & Co. Ltd. were suppliers to the aircraft industry.[9] Particularly in the early part of the war, every sinew was being strained to produce as many aircraft as possible with the country's survival depending on it.

At Black and Decker, Alonzo Decker's son, also called Alonzo , was interested in why they were getting so many repeat orders for electric drills. His concern was that the tools were failing in some way and he set out to discover what was happening. He soon found the answer: the employees were taking the drills home, and this gave him an idea.

His next task was to persuade his father that there was an untapped domestic market. He succeeded, and in 1946, once the war was over, the company started making drills specifically aimed at the do-it-yourself market. With this they claimed to have created the DIY revolution, though whether this was the chicken or the egg is difficult to untangle. However, there was definitely a market there, as the millionth drill came of the production line in 1950, and soon they began adding other tools such as circular saws to their range.

In Britain, S. Wolf and Co. saw a similar opportunity though they were also motivated to find other markets with the run down of their wartime work. In 1949, they produced the Wolf Cub drill, a low-cost unit for just under $15 (£5) aimed at the home market.[10] Available to go with it were a whole range of accessories such as sanding disks, polishers and grinding wheels. More complex kits could turn it into a fixed '"Pillar" drill, or even into a simple lathe. They were enormously successful, with sales rising to some 100,000 per annum by 1955.[11]

The end of the war was a time of hope in Britain, with expectations that things would get better. In terms of new things for the home it was fuelled by a number of exhibitions, notably the "Britain Can Make It" exhibition of 1946/47, the "British Industries Fairs" and the "Festival of Britain" in 1951. While these had no direct impact on the home, they set up an expectation of higher standards and of projects to improve the house. With material short and most of the available skilled labor absorbed by the reconstruction program, the householder's thoughts turned to doing it himself.

What he needed was knowledge, and here the fledgling medium of television was well prepared to help. Handyman W.P. Matthew was a regular contributor to the BBC's "Household Hints", and its successors "For the Housewife", "Workshop" and "About the Home" which ran from 1946 to his death in 1955.[12] After that, the handyman slots on "About the Home" and other DIY programs were taken over by Barry Bucknell who became so popular in the 1950s and 1960s that at its peak the program had seven million viewers, and the BBC needed ten staff to deal with the 40,000 letters a week he received.[13]

It was a simple matter for the "make do and mend" attitudes of the war years to metamorphose into do-it-yourself. The large number of new homeowners in the postwar building boom fed the trend, but there were also all those who had bought their houses during the 1920s and 1930s who were looking to update their homes to the latest clean lines that were now the fashion.

The "must-have" tool for all these budding DIYers was, of course, the electric drill with a set of basic accessories for sanding, polishing and sawing. Wolf was already in the market, but in 1954 they faced some competition when the British firm Bridges started making drills for the domestic market. In the same year, Black and Decker started selling into the UK, but technical problems held them back and allowed Bridges to get a foothold. The result of the competition was a reduction in Wolf's sales.

In 1958, Black and Decker attacked the British market more seriously with their D500 drill selling at just under $20 (£7), an attractive price at the time. First they test-marketed this with heavy advertising in the Midlands region before going nationwide. The campaign was very successful, with sales rising to 600,000 per annum by around 1961. The result was that they largely grabbed the market; Wolf and Bridges' shares diminished and they withdrew to concentrate on other sections of the power tool market.

Black and Decker knew their market worldwide, and also realized that constant innovation was the way to keep ahead. They were aware that the one drawback of the electric drill was that it has to be plugged in somewhere. There was a story of a storm window contractor who was going to make the instalation while the house owner was away. He had asked that the porch light be turned on so that he could use that socket to power his drill, but the owner forgot. The result was he was unable to do the job.[14] It was stories like that which led the search for a portable drill.

While rechargeable nickel cadmium batteries were available, it was only possible to cram four of them into the handle of the drill. This meant that there was only 4.8 V available instead of the 110 or 220 V of the mains. It required a radical redesign and the trade-off was that to obtain the same torque the drill had to run slower and hence take longer to drill a hole. That, however, seemed a reasonable compromise to achieve portability.

In 1961, the C600 appeared in the US at just under $60 instead of the usual $16 for a mains-powered drill. Despite the hefty increase in price, the boat builder and plumber, who might be standing in water at the time, for safety's sake needed one. Those working on roofs, or anywhere where it was inconvenient or impossible to have mains power leads dangling, also had to have one. As a result, the cordless drills sold well.

In Britain, the arrival of the two-speed drill in 1962 had the greatest effect as it soon replaced the single-speed unit. Even then, by the end of the decade the drill market was showing signs of saturation and Black and Decker, like its competitors, started to look elsewhere to continue its growth. A similar situation occurred in the US where as early as 1958 a survey found that almost three-quarters of handymen owned an electric drill.[15]

Black and Decker in particular put a lot of effort into reducing the costs of their motors and it then made sense to supply tools such as sanders with their own motor, the so-called integrated unit. This saved the bother for the user of having to fit the accessory every time it was required. By the late 1970s, the sector of integrated power tools was growing fast and was catching up with the drill market in value.[16] However, Black and Decker had attracted competition and their share of the market started to fall from around 90% to more like 60% as foreign companies such as Bosch, Hitachi, Makita and others started to make an impact on items other than the drills.

By this time, the DIY person was becoming more adventurous and could see the use for more tools. This led to the introduction of new products that were only really practical as separate units rather than accessories, including routers, planers and jig saws. Armed with these, the handyman could tackle jobs in his limited spare time that he wouldn't have considered before.

The weakest point was the cordless tools, but as battery technology improved this issue could be tackled. The first step was to make the battery removable which meant that another fully-charged pack could be inserted and work carry on. Otherwise, the tool became useless for the hours needed to recharge the battery. As long as the user also bought a spare battery pack then he could swap it into the tool and carry on working.

The next step was to increase the battery voltage and, hence, the power of the tool. This required the battery pack to be larger. With it usually slotting into the end of the handle, the logical thing to do was simply to make it larger to accommodate the increased cells. With a bit of careful design this didn't look too ugly and the weight could be balanced so that the drill was comfortable to hold and use. Now the cordless devices could largely match the corded ones for performance.

With all these developments, the sales of power tools rose and rose despite there being a few problems along the way (Fig. 22.2). The change to integrated tools began to drive the market and it really took off during the home ownership boom of the 1980s when the new owners wanted to upgrade them to modern standards. Thus power tools had become essentials for the home where DIY jobs were undertaken, and had changed the way that they were approached, making many more manageable.

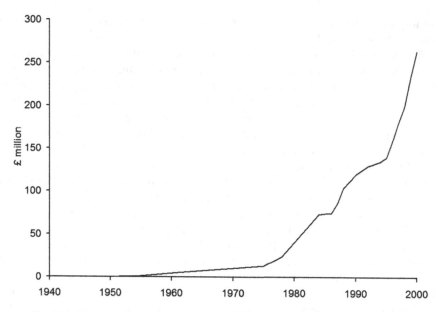

Fig. 22.2 The rise in UK sales of electric power tools, 1940–2000. Source: Author.[17]

Electric power came later to garden machines. The mower was invented in the nine-teenth century but it took until between the wars for electric motors to be applied to them. Ransomes, for example, produced a 16-in. electrically powered mower in 1926.[18] With so few houses having electricity, and that being rather expensive, it was hardly surprising that they weren't very popular.

Small petrol engines ruled, and if the user couldn't afford it, he simply had to push it. This was satisfactory for a very small lawn, but soon became tiring if a reasonable area was involved. With the rise of owner-occupied houses, each with their piece of grass, there was a considerable demand for something to keep it tidy.

It seemed logical to use electric motors instead of the petrol engines, but it took a very long time before they were produced. Instead, electricity was applied to other garden machines. In 1957, Black and Decker produced a portable electric hedge trimmer, the advantage being that it could be made lighter even though it required a trailing cable.[19] They went on in 1961 to make a cordless version using the technology developed for the electric drill, with the same rationale of portability and use in places where it was incon-venient or impossible to get mains power.

While the idea of the rotary mower had been around since before World War II, it was only in 1964 that Swedish inventor Karl Dahlman came up with one that hovered over the grass and was hence easy to move. He used the idea of Christopher Cockerell's Hovercraft by adding a fan above the cutting blade so that the mower floated off the ground. The first units used petrol engines, but by 1969 electric versions were also available.[20]

The electric motor, simpler and lighter than an engine, helped create a small compact mower suitable for the increasingly minuscule lawns in British gardens. While a push mower would have been adequate for most of them a powered mower was quicker, less work and had more kudos. Black and Decker also saw the opportunity and, though they had introduced mowers in the US a little before, brought out their small Lawnderette rotary mower in the UK. This used wheels as the Flymo principle was patented.

By the early 1970s, Qualcast, a traditional British mower manufacturer, joined the fray with their Concorde mower. It used the conventional horizontal spiral blades that had become the standard arrangement, but was scaled down and powered by an electric motor. It was light and easy to use, but unlike the Flymo and other rotary mowers at the time, it picked up the grass cuttings and left stripes on the lawn. It proved very popular and continued in production for more than 40 years.

By the 1980s, Flymo and Qualcast were locked in battle for the mower market, Flymo under the slogan, "Why slow-mo when you can Flymo?" and Qualcast hitting back with, "It's a lot less bovver than a hovver". It very much depended on the user's requirements and whether they were concerned about removing the clippings and producing stripes or not. Flymo eventually conceded some of the point by producing variants of their mowers which did collect the cuttings, and even some with wheels instead of hovering. On the other hand Qualcast also produced rotary mowers, showing that there were some advantages in each arrangement.

The lighter weight of the electric motors could also be an advantage in other existing garden products. Other companies followed Black and Decker's lead in making hedge trimmers suitable for the smaller suburban hedges. Also the traditional chain saw could be scaled down for occasional use and be powered by an electric motor. While its trailing cable meant it might not be suitable for climbing trees, it was useful for cutting up logs that had already been felled.

The use of small motors also brought innovation. The weed trimmer, or "strimmer", used a nylon filament rotating at high speed to cut rough weeds and grass around edges and trees where the mower couldn't reach, or for cutting the edges of lawns. It became the essential complement to the mower. Also used to tidy up the garden in the autumn was the leaf blower, taking the place of the rake or simply the wind which would usually deposit the leaves in a corner. To save time, it was felt worth having a special machine for just this task.

To reduce the bulk of garden waste the shredder came into its own. The electric ones were scaled-down versions of engine-powered commercial machines and hence suitable for the smaller garden. Another device appeared to enable the householder to clean all the paths and patios, or even the automobile. That was the pressure washer which used an electric motor-driven pump to deliver water at high pressure to a nozzle. The resulting jet was sufficiently strong to cut through dirt, leaving a clean surface.

What all these products had in common was that they used small, cheap electric motors. The mastery of these led to a whole range of devices that transformed do-it-yourself either in the house or in the garden. By the end of the century, virtually no one would attempt the tasks around the home or garden without the appropriate power tools. It meant that a greater range of household improvements could be achieved, or a more sophisticated garden could be managed.

Thus they became inextricably linked to the way people lived and the standards that they regarded as normal. In a world where more and more homes were owner-occupied, hiring labor to undertake the many tasks was either too expensive or unobtainable. Hence, homeowners had to do these things themselves. With limited time available, they were glad of all the help they could get and so these machines were a great boon and became essential to support the way people wanted to live.

NOTES

1. Scott, D. Electric drills history, available at: http://www.ehow.com /about_5041061_electric-drills-history.html.
2. The Tool Website. Electric drills, available at: http://goes.flexinet.com.au/images/ms/Electricdrills.pdf.
3. *Entrepreneur*. S. Duncan Black and Alonzo G. Decker Sr.: Power tool potentates, available at: http://www.entrepreneur.com/article/197612.
4. Black and Decker. History of Black+Decker, available at: http://www.blackanddecker.co.uk/about/history/.
5. Highbeam Business. Power-driven handtools, available at: http://business.highbeam.com/industry-reports/equipment/power-driven-handtools.
6. Cartwright, R. Power tools history, available at: http://ezinearticles.com/?Power-Tools-History&id=976405.
7. Lane, W, and =McCullough, P.M, Porter-Cable: 90 years of making power tools. *Journal of the International Academy for Case Studies*, January 1, 2000, also available at: http://www.the-freelibrary.com/Porter-Cable%3a+90+years+of+making+power+tools.-a0209043597.
8. S Wolf and Company, Catalogue, 1930.
9. Grace's Guide, S. Wolf and Co., available at: http://www.gracesguide.co.uk/S._Wolf_and_Co.
10. Ingenious. Power tools, available at: http://www.ingenious.org.uk/site.asp?s=RM&Param=1&SubParam=1&Content=1&ArticleID={2A0DB702-99B8-4AC8-82F2-98C9AF953447}&ArticleID2={BA51BAF2-335A-4C4D-8A0F-8E4F720AB2C6}&MenuLinkID={941FDB3F-21D2-4818-97CF-C3DE4E8AA560}.
11. Competition Commission. Black & Decker: A report on the course of conduct pursued by Black & Decker in relation to the supply of power tools and portable work-benches intended for domestic use. Chapter 2, available at: http://webarchive.nationalarchives.gov.uk/20111202195250/http://competition-commission.org.uk/rep_pub/reports/1989/258black_decker.htm.
12. *Radio Times* entries for the period, available at: http://genome.ch.bbc.co.uk/search/0/20?svc=9371533&q=W+P+Matthew.
13. Powell, H. Time, television and the decline of DIY. *Home Cultures*, 6:1, 89–108; Johnson, P. The rise and fall of the do-it-yourself bug. *The Telegraph*, October 24, 2014.
14. Fischetti, M.A. Case study: Cordless drill: Rejuvenating a product. *IEEE Spectrum*, 24:5, 32, 1987.
15. Gelber, S.M. Do-It-Yourself: Constructing, repairing and maintaining domestic masculinity. *American Quarterly*, 49:1, 66–112, 1997.
16. Price Commission. Prices, cost and margins in the manufacture and distribution of portable electric tools, February 21, 1979.
17. Data derived from the following with some calculations from the available data for some points: Price Commission, as above; Competition Commission, Black & Decker: A report on the course of conduct pursued by Black & Decker in relation to the supply of power tools and portable work-benches intended for domestic use. Chapter 2, available at: http://webarchive.nationalarchives.gov.uk/20111202195250/http://competition-commission.org.uk/rep_pub/reports/1989/258black_decker.htm; Bardsley, N. (Ed.), *Power Tools 1999 Market Report, Key Note*; Fenn, D. (Ed.), *Power Tools 2002 Market Report, Key Note*.
18. The Hall and Duck Trust, collectors of vintage lawnmowers. History of lawn mowers, part 7, available at: http://www.hdtrust.org/#/1930s-1940s/4573645043.
19. Black and Decker. About Black+Decker, available at: http://www.blackanddecker.co.uk/about/highlights/.
20. DIY in the era of the New Elizabethans. DIY Week Special Report, July 27, 2012.

23

The Electric Century

Study the past if you would define the future.

Confucius

On the evening of 13, 1977 a series of bolts of lightning hit high-voltage transmission lines in the area of New York.[1] Nine million people in the city were plunged into darkness. So began what Mayor Abraham Beame called a "night of terror" as the poor neighborhoods erupted into looting and arson. The rioters smashed their way into stores, 1600 of which were damaged, and made off with whatever they could carry: TV sets, drink, rifles, and diamonds. Even cars were hot-wired and driven away from a Bronx showroom.

One thousand fires were reported and some 3700 people were arrested for looting and rioting.[2] Neighborhoods from East Harlem to Bushwick were devastated, and the total cost was thought to exceed $300 million. In the event, it took over 24 h to get the whole city back on. It just showed how important the electricity supply was to modern society, and that law and order could break down without it.

Curiously, a much bigger event in 2003, when the whole of the Northeast US was blacked out, leaving some 50 million people without power, passed off peacefully. However, the chaos that otherwise ensued was quite enough to point up the problems. Modern society just can't function without an abundant supply of electricity.

While the first industrial revolution was mostly about changes in the way people worked, the coming of electricity had its greatest impact on the way people lived. It brought comfortable and convenient homes. Lighting easily extended the day, the control of heating and cooling brought a comfortable environment, while a multiplicity of appliances made everyday life so much easier.

Transport was also transformed, with electric trains and streetcars, and particularly electric subways which allowed people to live in the suburbs. The shape of cities was utterly altered with high-rise buildings made possibly by the electric elevator. To reach the below-ground trains or the upper floors of shopping malls, escalators were needed to cope with the numbers of people moving around a modern city.

© Springer International Publishing AG 2018
J.B. Williams, *The Electric Century*, Springer Praxis Books, DOI 10.1007/978-3-319-51155-9_23

Looking back, it can be seen how this castle was built; how one development enabled another and each took the technology a little further until the whole thing was constructed. This doesn't mean that progress was linear. One development enables others, but the route taken is often determined by other factors. We are dealing with human beings and their infinite variability. However, a general outline can be observed.

First came the exploitation of electricity. An important strand was electric lighting; this rapidly vanquished all other forms of lighting, and though it has developed in a number of ways it is still one of the mainstays of life today. Without it, the extension of the day that we take for granted would barely be possible. It is very difficult now to imagine life without light at the flick of a switch.

It also brought electrical power which led to streetcars and high-speed trains. Then it powered the factories, and soon crept into the home in myriad gadgets and appliances. Life without refrigerators, freezers and washing machines is unthinkable. They are now regarded as necessities, and there are many more items such as vacuum cleaners and dishwashers that could be added to that category.

Electricity is a huge subject and in this book it has only been possible to skim through just to show its impact. However, another huge area, electronics and its offshoots of computing and telecommunications, has had to be left to another book, *The Electronics Revolution*. There simply wasn't room here.

Obviously, it has not been possible to include everything and the selection has been of things felt to have the greater impact on everyday life. It is a set of personal choices and so will not please everybody, but hopefully it achieves its objective of tracing the main outlines and showing how these technologies have piled on each other and the impact they have had on people's lives.

What of the future? This is always a hazardous area and many fall into the trap of not realizing how long the machines that we already have will last. In 10 years' time most of the devices will still be the same. However, we can see some trends. In developed countries the continuous rise in the consumption of electricity has largely leveled out. Partly this is due to economic conditions, but it is also that there has been a deliberate policy to reduce consumption driven by concerns about climate change.

In electricity generation we see the trend towards greater reliance on renewable sources. This is a curious word for something that ultimately comes from the sun or the spinning of the Earth. Some other languages use a word more like "enduring" which seems to fit the bill rather better. Windmills are well advanced and solar panels are beginning to make an impact. However, the current thinking is still for there to be a mix of various sources; it would appear difficult to obtain all that we need from just renewables.

Concerns about the impact of hydroelectric schemes have led to a trend away from them because of the environmental trade-offs necessary to implement the dam and large lake. Politicians have always been keen on impressive projects, but do they have to be so large? When modern wind power was first mooted, the critics said it was ridiculous because it would require thousands of wind turbines. That was in an age of relatively few large power stations, where efficiency was a matter of size. It required a change of mindset to realize that it was perfectly possible to have those huge numbers of windmills. So why not implement many small hydroelectric schemes, particularly ones that work with the flow of a river and don't require much in the way of a dam? After all, electricity is a convenient way of distributing the power of flowing or falling water away from the source.

Nuclear power was once projected as the answer, but though it has served us well, a series of accidents has turned people against it. Three Mile Island, Chernobyl and Fukushima cannot be wished away. Another factor is that they are very expensive to build, though the fuel is cheap by comparison. Governments with their emphasis on economy are not going to put up the initial cost, and private industry looks for easier targets. It seems likely these fission reactors are on the wane.

On the other hand, research on fusion is still progressing with conditions close to what is required being reached. A power station using this, when practical, is likely to be even more expensive to build. If so, it would run into exactly the same problems. We would need to be very desperate for sources of energy before anyone would to commit to such expensive things. In addition, the future of supply is thus somewhat uncertain.

In terms of consumption we have seen the trend to more efficient lighting and appliances. Incandescent lamps have been replaced by low-energy bulbs and LEDs. Refrigerators and freezers have better insulation and so use less power. Other appliances have increased in efficiency. Buildings are better insulated. While these trends will probably continue, and the odd robot device appear, the changes are likely to be slow and evolutionary. Almost by definition, revolutionary changes are nearly impossible to predict.

As the coming of electricity has clearly had such an impact, the question is often raised as to whether technology drives history. From Karl Marx onwards there have been proponents of the concept of technological determinism. If a particular technological development occurs, is the result a foregone conclusion? Marx is quoted as saying: "The hand mill gives you society with a feudal lord; the steam mill, society with the industrial capitalist".[3]

One does wonder about some of the commentators whether they have ever shaped metal with a file or wielded a soldering iron or, more particularly were ever involved in the process of producing a new technological product. If so, they would know that nothing is certain about the result. Everything can be there that seems to be exactly what is required and then it is a complete flop. On the other hand, something that seems quite improbable can be a roaring success.

Why does virtually every household have a clothes washing machine but far fewer have a dishwasher? Why did it take around a century for a quarter of homes in Britain to have a telephone, and then suddenly accelerate and in a few years become almost universal? Why did the mobile cell phone take off much more rapidly? The answers to these questions don't lie in technological developments but in the way people respond to them.

A technological development does not dictate an outcome, it is an enabler; it sets up opportunities. There is no certainty which of these will be taken. Sometimes there is an immense length of time before interest grows, maybe because the products aren't quite right, or simply the idea's time hasn't come. Many appliances have taken half a century from their invention to being a staple in the home.

Looking at the change in people's lives is also interesting. Few at the end of the century would want to return to conditions at the beginning, without electric light and the convenience and comfort of central heating or air conditioning. They are also much too attached to their washing machines and other appliances. No one can imagine washing with heated coppers and dolly tubs.

Transport is utterly changed, with electric trains, cars, and airplanes which wouldn't function without their electrical equipment. No one now can imagine depending on horses for transport, either riding them or using them to pull a vehicle of some sort. It is difficult now to appreciate the usual slow speeds and dirt of a steam train. This increased mobility has shaped cities; they cannot be so large, with their sprawling suburbs, without the transport to take people from their homes to their work or other facilities.

The place of work has greatly altered with it being far cleaner, warmer and lighter. Where in the nineteenth century steam was beginning to replace muscle power, the flexibility of the electric motor took this much further. Automation increasingly replaced manual work, decreasing the number of repetitive jobs, and largely eliminating the dangerous ones. The type of jobs that people do is almost completely different in the course of the century.

All of these things depend in some way or another on the development of electricity and its children of electronics, telecommunications and computing. It is these that have shaped almost all the changes in the conditions of life in the course of the century. While the nineteenth is usually thought of as the century of coal and steam, undoubtedly the twentieth is the electric century.

NOTES

1. Greenberg, D. Where have all the looters gone? *Slate,* August 15, 2003, available at: http://www.slate.com/articles/news_and_politics/history_lesson/2003/08/where_have_all_the_looters_gone.html.
2. Chan, S. Remembering the '77 blackout. City Room, available at: http://cityroom.blogs.nytimes.com/2007/07/09/remembering-the-77-blackout/comment-page-7/?_r=0.
3. Smith, M.R. and Marx, L. *Does Technology Drive History?: The Dilemma of Technological Determinism*, p. 123.

Bibliography

Addison, P. (2010). *No turning back: The peacetime revolutions of post-war Britain.* Oxford: Oxford University Press.

Addison, P. (1995). *Now the war is over: A social history of Britain 1945–51.* London: Pimlico.

Arthur, M. (2007). *Lost voices of the Edwardians.* London: Harper Perennial.

Atherton, W. A. (1984). *From compass to computer: A history of electrical and electronics engineering.* London: Macmillan.

Bardsley, N. (Ed.). (1999). *Power tools 1999 market report.* Hampton: Key Note.

Batchelor, R. (1994). *Henry Ford: Mass production, modernism and design.* Manchester: Manchester University Press.

Bennett, C. N. (1913). *Handbook of kinematography: The history, theory, and practice of motion photography and projection.* London: Kinematograph Weekly.

Birch, J. (1799). A letter on the subject of medical electricity. In G. Adams (Ed.), *An essay on electricity* (5th ed.). London: Dillon.

Bogdanis, D. (2005). *Electric universe: The shocking true story of electricity.* New York: Crown.

Bowen, H. G. (1951). *The Edison effect.* West Orange, NJ: The Thomas Alva Edison Foundation.

Bowers, B. (1982). *A history of electric light and power.* London: Peregrinus/Science Museum.

Bowers, B. (1998). *Lengthening the day: A history of lighting technology.* Oxford: Oxford University Press.

Byatt, I. C. R. (1979). *The British electrical industry, 1875–1914: The economic returns to a new technology.* Oxford: Clarendon Press.

Byers, A. (1988). *Willing servants: A history of electricity in the home.* London: The Electricity Council.

Clout, H. (2007). *The times history of London.* London: Times Books.

Contact. (1917). *An airman's outings.* Edinburgh: Blackwood.

Coursey, P. R. (1919). *Telephony without wires.* London: The Wireless Press.

Croy, H. (1918). *How motion pictures are made.* New York: Harper & Brothers.

Derry, T. K., & Williams, T. I. (1960). *A short history of technology.* Oxford: Oxford University Press.

Dimbleby, D. (2007). *How we built Britain.* London: Bloomsbury.

© Springer International Publishing AG 2018
J.B. Williams, *The Electric Century*, Springer Praxis Books, DOI 10.1007/978-3-319-51155-9

Euromonitor. (1986). *The small electrical appliance report*. London: Euromonitor.

Fenn, D. (Ed.). (2002). *Power tools 2002 market report*. Hampton: Key Note.

Fessenden, H. M. (1940). *Fessenden, builder of tomorrows*. New York: Coward-McCann.

Fleming, J. A. (1924). *The thermionic valve and its developments in radiotelegraphy and telephony*. London: Iliffe & Sons.

Garfield, S. (2004). *Our hidden lives*. London: Ebury Press.

Gooday, G. (2008). *Domesticating electricity: Expertise, uncertainty and gender, 1880–1914*. London: Pickering and Chatto.

Gordon, B. (1984). *Early electrical appliances*. Shire: Princes Risborough.

Green, M. (2007). *The nearly men: A chronicle of scientific failure*. Stroud: Tempus.

Hammond, R. (1884). *The electric light in our homes*. London: Warne.

Hannah, L. (1979). *Electricity before nationalization*. London: Macmillan Press.

Hannah, L. (1982). *Engineers, managers and politicians*. London: Macmillan Press.

Hardyment, C. (1988). *From mangle to microwave*. Cambridge: Polity Press.

Harrison, B. (2011). *Seeking a role: The United Kingdom 1951–1970*. Oxford: Oxford University Press.

Hartley, R. J. (2008). *North London Trams*. Harrow: Capital Transport.

Harvey, C., & Press, J. (1989). *Sir George White of Bristol, 1854–1916*. Bristol: Bristol Branch of the Historical Association.

Hattersley, R. (2004). *The Edwardians*. London: Little, Brown.

Hennessey, R. A. S. (1972). *The electric revolution*. Newcastle upon Tyne: Oriel.

Hiley, N. (1998). At the picture palace: The British cinema audience, 1895–1920. In J. Fullerton (Ed.), *Celebrating 1895: The centenary of cinema*. London: J. Libbey.

Hiley, N. (2002). Nothing more than a craze: Cinema building in Britain from 1909 to 1914. In A. Higson (Ed.), *Young and innocent: The cinema in Britain, 1896–1930*. Exeter: University of Exeter Press.

Hounshell, D. A. (1985). *From the American system to mass production, 1800–1932: The development of manufacturing technology in the United States*. Urbana, IL: University of Illinois Press.

Howgrave-Graham, R. P. (1907). *Wireless telegraphy for amateurs*. London: Percival Marshall.

Hughes, T. P. (1983). *Networks of power: Electrification in western society, 1880–1930*. Baltimore: Johns Hopkins University Press.

Hunt, T. (2005). *Building Jerusalem*. London: Phoenix.

Issacs, G. A. (1931). *The story of the newspaper printing press*. London: Co-operative Printing Society.

Kynaston, D. (2007a). *A world to build*. London: Bloomsbury.

Kynaston, D. (2007b). *Smoke in the valley*. London: Bloomsbury.

Kynaston, D. (2009). *Family Britain 1951–57*. London: Bloomsbury.

Kinematograph. (1913). *Kinematograph year books*. London: Kinematograph.

Maddison, A. (1991). *Dynamic forces in capitalist development*. Oxford: Oxford University Press.

Marr, A. (2009). *A history of modern Britain*. London: Macmillan.

Mitchell, B. R. (1988). *British historical statistics*. Cambridge: Cambridge University Press.

Nevins, A. (1954). *Ford: The times, the man, the company.* New York: Scribner.

Oxford Dictionary of National Biography, online edition.

Parsons, R. H. (1940). *The early days of the power station industry.* Cambridge: Cambridge University Press.

Pimlott, B. (1995). *Hugh Dalton.* London: HarperCollins.

Political & Economic Planning (PEP). (1945). *Report on the market for household appliances.* London: PEP.

Poulter, J. D. (1986). *An early history of electricity supply.* London: Peter Peregrinus.

Price, C. (2007). *Depression and recovery.* In F. Carnevali & J.-M. Strange (Eds.), *20th century Britain.* Harlow: Pearson Education.

Pugh, M. (2008). *We danced all night.* London: Bodley Head.

Rogers, E. M. (1983). *Diffusion of innovations.* London: Collier Macmillan.

Scott, P. (2007). Consumption, consumer credit and the diffusion of consumer durables. In F. Carnevali & J.-M. Strange (Eds.), *20th century Britain.* Pearson Education: Harlow.

Smith, M. R., & Marx, L. (1994). *Does technology drive history? The dilemma of technological determinism.* Cambridge, MA: MIT Press.

Snodgrass, E. M. (2004). *Encyclopedia of kitchen history.* New York: Fitzroy Dearborn.

Stratosch, G. R. (1983). *Vertical transportation, elevators and escalators.* New York: Wiley.

Swan, M. E., & Swan, K. R. (1929). *Sir Joseph Wilson Swan.* London: Ernest Benn.

Taylor, A. J. P. (1977). *English history 1914–1945.* Oxford: Oxford University Press.

Tobey, R. C. (1996). *Technology as freedom: The new deal and the electrical modernization of the American home.* Berkeley, CA: University of California Press.

Uglow, J. (2002). *The Lunar men.* London: Faber & Faber.

Wheen, A. (2011). *Dot-dash to dot.com: How modern telecommunications evolved from the telegraph to the internet.* New York: Springer.

Weightman, G. (2011). *Children of light: How electrification changed Britain forever.* London: Atlantic.

Williams, R. (1992). *The long revolution.* London: Hogarth.

Wood, L. (1986). *British films 1927–1939.* London: BFI.

Index

© Springer International Publishing AG 2018
J.B. Williams, *The Electric Century*, Springer Praxis Books, DOI 10.1007/978-3-319-51155-9

Printed in the United States
By Bookmasters